Eliza Brightwen

Rambles with Nature Students

Eliza Brightwen

Rambles with Nature Students

ISBN/EAN: 9783337025458

Printed in Europe, USA, Canada, Australia, Japan

Cover: Foto ©berggeist007 / pixelio.de

More available books at **www.hansebooks.com**

Rambles
with
Nature Students

Photographed by]   *[JAMES LEVERSUCH.*
IN 'THE GROVE' GARDEN.

Rambles

with

Nature Students

BY

Mrs. BRIGHTWEN, F.E.S.

AUTHOR OF
'WILD NATURE WON BY KINDNESS' 'SIDELIGHTS ON THE BIBLE,' ETC.

ILLUSTRATED BY THEO. CARRERAS

SECOND IMPRESSION

LONDON
THE RELIGIOUS TRACT SOCIETY
4 BOUVERIE STREET, AND 65 ST. PAUL'S CHURCHYARD

Preface

IT is quite possible to take a walk in the country, through the most beautiful scenery, in lovely weather, with everything to conduce to our enjoyment and invigoration of spirit, and yet to return feeling bored and weary, and half inclined to say how dull the country is! That is one side of the picture.

On the other hand I have known young people come back from a ramble in a quiet and rather unpromising country lane, their faces beaming with pleasure, and their hands filled with an odd collection of specimens, leaves, mosses, stones, anything in fact which had taken their fancy as curious or interesting. Then eager questions are poured forth with bewildering rapidity, and it is easy to see that keen enjoyment has been derived from even this commonplace little stroll. May I point out that the difference between these two results simply arose from acquiring or not acquiring the habit of seeing intelligently what lies around us? If we pass everything by with our mental eyes shut, our physical eyes observe nothing.

I am going to take for granted that a large number of my readers belong to the former class, that they are intelligent observers, and yet are in need of a guide to help them to understand the thousand and one things that they may see in a country walk.

The curious objects in hedges, trees, and fields all have a purpose and a meaning, but very often these need interpretation for those who never have had the opportunity of acquiring facts in natural history.

The practice of putting down the results of each day's ramble, making notes of things seen or obtained, the first appearances of birds and insects, the flowering of trees and plants, will result in the course of a few months in a record possessing a certain value. We can thus compare one year with another, and note the differences in each, and the effect of temperature in hastening or retarding the appearance of flowers and insects, and the arrival of migratory birds.

The remarks I shall endeavour to make upon all these and other points will be the result of my own actual observations, made from day to day and noted down at once, so that any readers who may like to follow this plan can do so with ease, if they happen to live in the country or have access to it from time to time.

The first appearances of birds, insects, and flowers may vary somewhat as to date, according to the mildness or severity of the winter, so that I cannot promise that every object that I write about will be found upon the same day in the following year, but probably within a short period, earlier or later, each object will be discovered.

It need not be thought that one must be far away from cities in order to learn about nature. I live only twelve miles from Charing Cross, and yet I find abundant subjects for study in my own place and the adjacent common. I think it is especially interesting to try and find treasures in most unlikely localities.

Having on one occasion to wait a whole hour on a pouring wet day at Bedford railway station, I determined to see if I could collect anything to while away the time. Things looked very unpromising outside the station; new houses in the act of being built, heaps of sand and mortar, and plenty of mud everywhere, seemed hopeless enough; but a bare patch of common, across which ran a newly gravelled road, caught my eye; there might be possibilities in the gravel, so I made my way to the new road, and before long I had the pleasure of finding there several rare fossils, pieces of chalcedony and jasper, a shell impression, and sundry other treasures; so, in spite of rain and wind, my waiting hour passed, not only without weariness, but in positive enjoyment.

I believe I have heard of as many as fifty species of wild flowers being found in a single field, and a well-known scientist discovered an equal number of wild plants in a piece of waste ground in the outskirts of a large town.

Preface

It is a little discouraging to begin our natural history diary in January, just when all animal and plant life seem asleep for the winter; but perhaps we shall find to our surprise that there is hardly a day in the blankest season of the year, which will not afford us some sources of interest and much that will lead to pleasant thought and study. The limits of space will not admit of a daily ramble, and bad weather sometimes hinders outdoor study; so, for a little variety, I have sometimes discoursed upon objects taken from my own museum.

Contents

	PAGE
January	17
February	29
March	47
April	67
May	81
June	97
July	115
August	131
September	145
October	163
November	183
December	199

List of Illustrations

	PAGE
IN 'THE GROVE' GARDEN	*Frontispiece*
PTEROMALUS	20
OWL PELLET	21
HORSE-CHESTNUT TWIG	22
WITCHES' BROOMS ON BIRCH TREE	24
RESONANT FLINT	25
VENTRICULITE IN FLINT	25
VENTRICULITE IN CHALK	25
SCOLYTUS BORINGS	26
CLOTH MOTH	31
FUR AND FEATHER MOTH	31
CORK MOTH	31
MOTH LARVÆ IN PLUSH CLOTH	32
SNOW CRYSTALS	33
FROSTED LAUREL LEAF	34
SKELETON BULB	35
FOOTPRINTS IN SNOW	37
LESSER CELANDINE	39
HAZEL CATKIN	41
ALDER CATKIN	42
WASPS	44
AUCUBA BERRIES	49
MEALWORM BEETLE	51
WHITLOW GRASS	53
RUE-LEAVED SAXIFRAGE	53
DANDELION LEAF	55
NUTHATCH AND NEST	56
SYCAMORE FRUIT	58
MAPLE FRUIT	58
HORNBEAM SAMARA	59
HORNBEAM SAMARAS RIPE	59
ARAUCARIA SEED	59
WHITE POPLAR (MALE CATKIN)	60

List of Illustrations

	PAGE
White Poplar (female catkin)	61
Male Flowers of Yew	62
Yew Berries	63
Sallow Catkins	64
St. Mark's Fly	70
Death's Head Moth and Larva	72
Humble-bee Fly	75
Flowers of Ash	76
Birch Catkin	77
Sycamore Seedlings	78
Shepherd's Purse	80
Rhododendron	83
Gladiolus	84
Larder Fly	86
Bluebottle	86
Canary Grass	87
The Trinity Flower	89
Larch Blossom	91
Beech Catkins	92
Ivy-leaved Toad-Flax	93
Leaf-cutter Bee and Rose Leaf	100
The Hoverer-fly	103
Rhyssa Persuasoria	104
Flax	107
Snake-fly and Larva	109
Diagram of Growing Broad Bean	111
Podophyllum Bud	112
Sanguinaria	112
Flies in Amber	118
Polished Quartzite Pebble	120
Cornish Granite	121
Teasel	123
Teasel Head	124
Wild Succory	125
Dragon-fly Pupa	128
Dytiscus Marginalis and Larva	129
The Great Green Grasshopper	134
Bird's-foot Trefoil	135
Coprinus Comatus	137
Common Blue Butterflies reposing on Grass	138
Rhododendron	140
Egyptian Hieroglyph for Rejoicing	140
Palestine Oak	143
Young Blue-tit	148

List of Illustrations

	PAGE
LONG-EARED BATS	150
AGARICUS CAMPESTRIS	154
BOLETUS EDULIS	155
HYDNUM REPANDUM	156
CLAVARIA	157
ORIGIN OF CORINTHIAN ORDER	158
ACANTHUS MOLLIS	159
OROBANCHE SPECIOSA	161
PAPYRUS SYRIACUS	166
FLOWER OF PAPYRUS	167
PAPYRUS PAPER	168
WILD-ROSE GALL	170
OAK-LEAF GALLS	170
OAK-FLOWERS—GALLS	171
TURKEY OAK	172
COMMON ENGLISH OAK	173
LONG-STALKED OAK	174
SESSILE OAK	175
CEDAR CATKINS	177
YELLOW IRIS CAPSULES	179
DATURA	180
COLUMBINE	181
PIMPERNEL	181
SILVER-TREE SEED	182
CYCLAMEN CAPSULES	182
OWL'S FOOT	186
JACANA'S FOOT	187
SNOW-SHOE	188
FOOT OF PTARMIGAN	188
DUCK'S FOOT	189
TRUMPETER PIGEON'S FOOT	189
FEET OF LARK	189
FLY KILLED BY FUNGUS	190
CHAMBERED NAUTILUS	191
ARGONAUT, OR PAPER NAUTILUS	193
HELIX LAPICIDA AND MEDICAGO HELIX	194
CATERPILLAR SEED	194
NEST OF BEE IN SNAIL SHELL	195
LEAF MINED BY MOTH LARVÆ	197
THE HOLLY	202
OTOLITHS	203
AN OTOLITH SCREEN	204
PALM SCALE	206
COCHINEAL INSECT ON CACTUS	207

List of Illustrations

	PAGE
APPLE MUSSEL SCALE	207
LEPISMÆ SACCHARINA	210
PURPLE HELLEBORE	212
SPINNING MITE	213
BEETLE MITE	213
CHEESE MITE	213
OWL'S SKULL	216
DUCK'S SKULL	216
WOODCOCK'S SKULL	217
PHEASANT'S SKULL	217

January

> 'To him who in the love of Nature holds
> Communion with her visible forms, she speaks
> A various language; for his gayer hours
> She has a voice of gladness, and a smile
> And eloquence of beauty, and she glides
> Into his darker musings with a mild
> And healing sympathy, that steals away
> Their sharpness, ere he is aware.'
>
> Bryant.

Rambles with Nature Students

January

> ' My heart is awed within me, when I think
> Of the great miracle that still goes on
> In silence round me—the perpetual work
> Of Thy creation, finished, yet renewed
> For ever. Written on Thy works, I read
> The lesson of Thine own eternity.'
>
> Bryant.

PTEROMALUS

ALTHOUGH the weather is very cold, I see a quantity of little hardy flies upon the window-pane. Apparently they are unaffected by a temperature which paralyses almost all other insects in the depth of winter.

This special little fly, *Pteromalus*, has a very curious life-history ; for it lays its eggs in living caterpillars, chrysalides, or hybernating bluebottle flies. The eggs hatch into very minute grubs, which feed upon, but do not kill, the unfortunate insect until they are full grown, when they emerge from the creature they have preyed upon, turn into tiny chrysalides, and in due time appear as perfect flies. They are so excessively small that they can creep through a mere crevice at the back of a picture-frame and make their way under the glass. Thus I have frequently found thirty or forty of them spread over the inner surface of some valuable print, and there was no

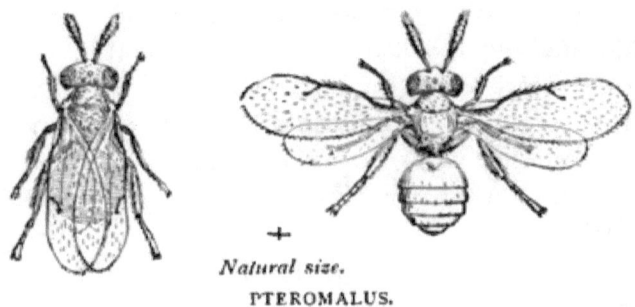

Natural size.
PTEROMALUS.

getting at them but by removing the picture and reframing it.

These flies perform a very useful office in reducing the number of caterpillars and other noxious insects which would otherwise abound in our gardens.

Owls

We are constantly hearing the brown owl's hoot, both in the daytime and in the dusk, and occasionally I see it and the white barn owl flitting across the lawn in the twilight.

These birds are of essential value in ridding the land of mice; they are like winged cats always on the watch for their prey, and very successful they are in catching, not only mice but young rats, sparrows, and beetles.

Owls like to roost on certain trees which afford them a thick covert during the day, and beneath those trees I often find large grey pellets, consisting of the fur and bones of rats and mice, which it is the habit of the owls, as they cannot digest them, to reject each morning after their nightly feast. When owls are kept as pets, their raw meat diet should include a mixture of small feathers, or fur of some kind, else the birds will not continue in a healthy state. The frequent occurrence in their pellets of the wing cases of the dark-blue dung-beetle shows that this is a favourite article of diet with the owls.

In order to ascertain the number of mice and other rodents destroyed by these useful birds, seven hundred and six pellets of the barn owl were carefully examined, and in them were found the remains of sixteen bats, three rats, two hundred and ninety-three voles or field mice, one thousand five hundred and ninety shrews and twenty-two small birds. We thus see that without their aid the farmer would find it very difficult to save his crops from devastation, and that these useful birds should be protected and encouraged by every means in our power.

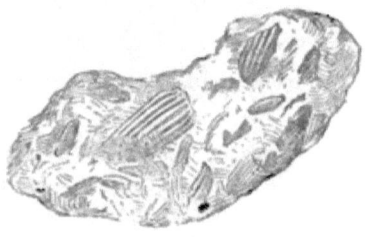

OWL PELLET.

A few years ago, when the crops in Southern Scotland were threatened with complete destruction by field mice or voles, great flocks of owls appeared on the scene, and corrected a plague which human science had proved quite unable to deal with.

Leaf-Scars

Now that the trees are leafless, we can readily observe the marks upon the branches called leaf-scars, which show where leaves have been.

Some trees, such as the sycamore, the wayfaring tree, and others, have opposite leaves; others produce them alternately or at varying distances and in a variety of ways; the study of leaf position is known in botany as Phyllotaxis, and it is to the individual differences in bud-growth that we owe much of the beauty of our woods.

Each tree has branches varying in form, in light-

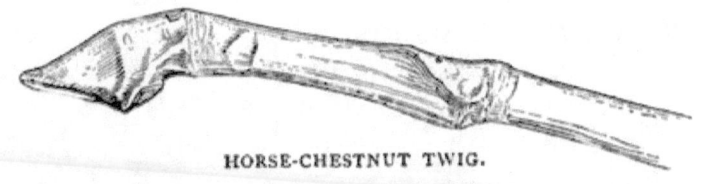

HORSE-CHESTNUT TWIG.

ness and density, and hence arises the exquisite play of light and shade which we cannot fail to admire when trees are grouped together.

One curious fact about the horse-chestnut may easily be noted at this season. Amongst the smaller branching twigs some may be found which are almost exact counterparts of a horse's foot and leg. As shown in the illustration, there are the hoof and six or eight nail marks of the shoe, the fetlock joint and part of the leg.

According to the angle at which the twig is growing will depend its resemblance to a fore or hind leg. There appear to be three suggested derivations of the name of this tree. The word 'horse' is a common prefix denoting anything large or coarse, such as horse-mushroom, horse-radish, horse-parsley; and so it may have been applied to this tree, which grows vigorously and has large leaves. One writer, however, explains the name as being a corruption of the word 'harsh,' as the horse-chestnut fruit is harsh and austere, compared with the sweet chestnut with its eatable nuts. There remains the third derivation, arising from the curious mimicry we find in the twigs and branches, which seem to be quite a likely reason for bestowing a name alluding to the fact.

A little ingenuity in neatly cutting and trimming the mimic horse's leg will result in a woodland curio which will surprise those who have never happened to notice the shapes which horse-chestnut twigs assume.

HORNBEAM

Some Hornbeam trees are attacked by a kind of parasitic fungus (*Exoascus carpini*), which so seriously interferes with the flow of the sap that a multitude of small interlacing shoots are the result. These give to the tree in winter the effect of being laden with birds' nests.

Each year these tufts increase in size, until the branches become weighed down with their unnatural burden. These 'witches' brooms,' as they are popularly called, occur also upon the birch and several species of pines, larches, and spruce firs.

It is still, I believe, a moot question whether these unusual growths may not be the work of a

WITCHES' BROOMS ON BIRCH TREE.

gall fly instead of a fungus; and here is a field for the ingenuity of a young observer to exercise itself upon.

VENTRICULITES

Those who have access to a chalk-pit may like to know that the long slender flints so often to be found there are singularly resonant.

If two flints are attached by a piece of string and struck against each other whilst held suspended in the air, they emit a sweet ringing sound almost like that of a bell.

Certain fossil sponges called ventriculites may also be found amongst chalk *débris*; they are usually met with in two pieces, having snapped asunder at the narrowest part; but by putting the upper and lower halves together we may easily imagine how they looked when growing on some sea-shore countless ages ago.

I pick up flint ventriculites in my garden amongst other

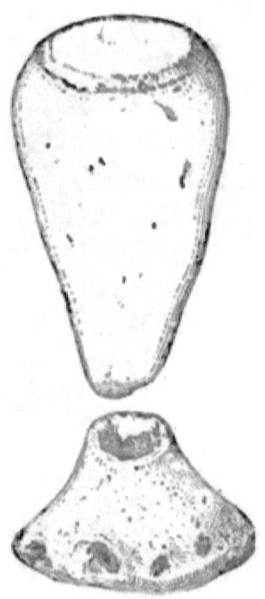

VENTRICULITE IN FLINT.

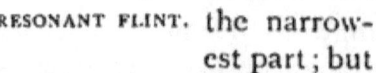

RESONANT FLINT.

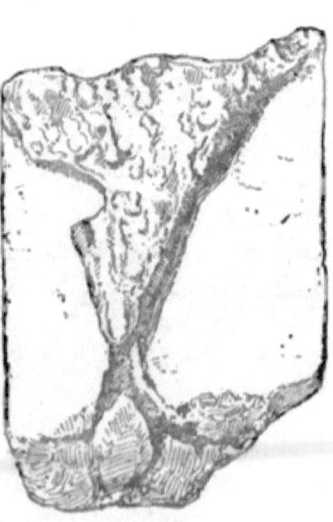

VENTRICULITE IN CHALK.

stones lying on the surface of the ground, and this fact, taken in connection with the existence of our rounded pebbles, shows that in early days the sea must have rolled over this part of Middlesex, although now it is the highest ground all round London.

Another proof of this fact was afforded by our finding a fossil crab, which was discovered about ten feet below the surface by some men who were digging a well.

ASH-BARK BEETLE (*Scolytus*)

A piece of bark has fallen off an old gate-post, and has revealed some markings on the wood beneath. These I find are the work of a small beetle, which

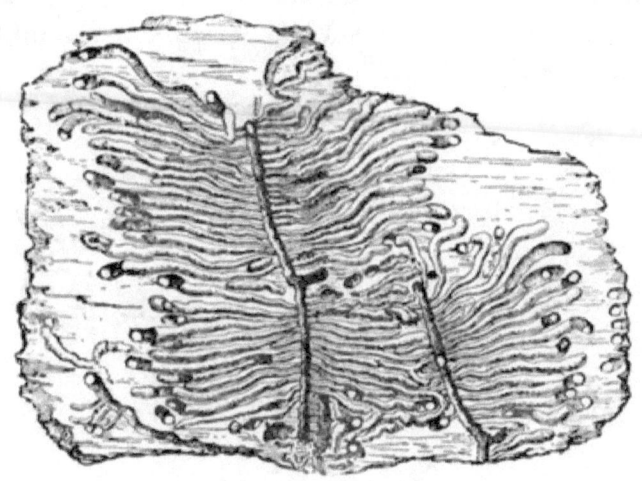

SCOLYTUS BORINGS.

burrows under the bark of the ash tree and there lays her eggs. When the grubs come out each one lives and works in its own little tunnel, eating the wood as it goes along until it is full grown and changes into a pupa and eventually into a perfect beetle, when it gnaws its way out, leaving a small round hole at the end of the tunnel.

An allied species does grievous damage to elm timber; whole forests are sometimes destroyed by this apparently insignificant insect.

The beetle bores into the tree-stem, makes a central gallery, and from it she bores small side galleries with wonderful regularity side by side, and at the end of each of these alleys she lays an egg; and when the larvæ are hatched they gnaw the wood in a straight line, always enlarging the gallery as they themselves grow bigger, so that the result upon the wood is a curiously symmetrical pattern.

Other beetles make curved galleries of intricate design, of which I have several specimens resembling delicate wood-carvings.

February

'The sunbeams on the hedges lie,
The south wind murmurs summer-soft;
The maids hang out white clothes to dry,
Around the elder-skirted croft.
A calm of pleasure listens round,
While fancy dreams of summer sound,
And quiet rapture fills the eye.'

Clare.

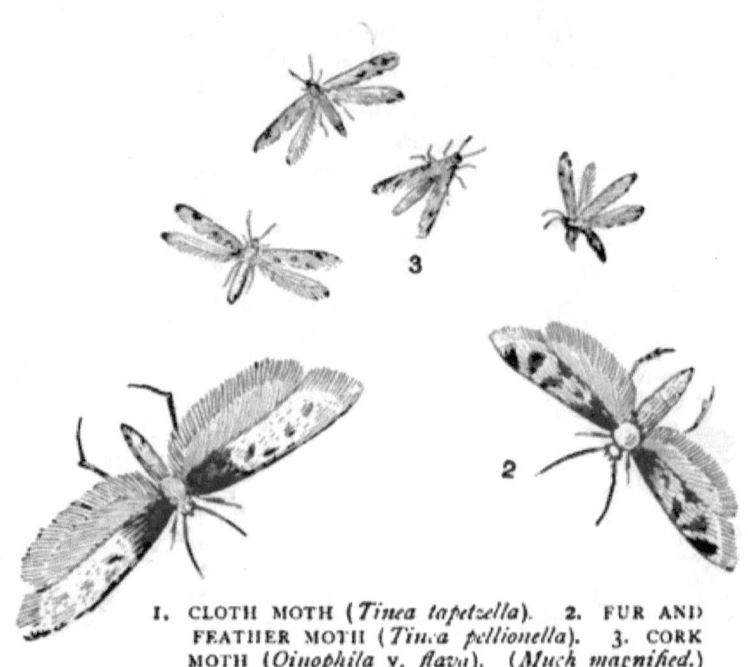

1. CLOTH MOTH (*Tinea tapetzella*). 2. FUR AND FEATHER MOTH (*Tinea pellionella*). 3. CORK MOTH (*Oinophila* v. *flava*). (*Much magnified.*)

February

MOTHS

DURING the past few years I have made the acquaintance of a great many members of the moth family.

A small room built out of the conservatory was found to be too damp for my daily use, and was for a while unused in consequence. I find that the moths have been having grand times there; they found out some boxes of curious feathers, and reduced them to shreds and atoms; they reared extensive families in the buffalo skin which carpets the floor; a stuffed gazelle has afforded a delightful feeding-ground for another species. I find that a box of owl pellets is swarming with *Tineas*; in fact, nothing seems to have come amiss to these little plagues. They can adapt themselves to digest every kind of material, and very diligently do they set to work to reduce feathers, cloth, furs, and stuffed animals and birds to a heap of dusty fragments. One is familiar with the ordinary moth cases containing the grubs, and sometimes the small white larvæ make tunnels in the substance they are devouring, but in the room I speak of a certain red plush table-cover contained a number of neat

MOTH LARVÆ IN PLUSH CLOTH.

little oval cells, and in each reposed a fat white grub, no doubt the maker of the cell. Since the cloth is ruined, I have allowed these innocent babies to remain in their cradles, and I shall watch them turn into chrysalides, and eventually into moths.

The smallest specimens of this destructive tribe that I have yet met with are the cork moths; they lay their eggs in the corks of old sweet wine, with the result that the grubs bore holes into the said corks, and thus let in the outer air and turn the wine into vinegar, and in this way thousands of bottles of

February

choice wine become spoiled by an enemy so minute as to be very seldom seen in the winged state.

This fact points to the necessity of protecting the corks of valuable wine by sealing-wax or metal capsules.

Snow Crystals

Winter does not afford many living creatures as subjects for our study; we must therefore turn our attention to other natural objects. To-day, as snow is falling, we will go out with a powerful magnifying glass and examine the beauty of snow crystals.

It is not always possible to see the formation of snow; if there is much wind, the crystals are apt to be broken, and unless the cold is severe the flakes melt away too soon to allow us to examine them. In sharp frost, on a calm day, the first flake of snow we look at through a lens will reveal the beautiful six-rayed crystals of which it is composed, and although each one has invariably six points, yet the ornamentation is infinitely varied. Each lovely star is fringed with most delicate tracery, and the flower-like forms glisten like burnished silver.

SNOW CRYSTALS.

I have read somewhere that no fewer than a thousand different patterns and devices have been found of these snow crystals, and as we examine the flake we have placed beneath the glass, we see for ourselves something of the indescribable beauty of these 'ice-morsels.' The silvery frost-work upon the window-pane shows the same crystalline law, only the stars are often merged into continuous tracery, so that the six rays are not always so easily discerned as in the snowflake.

Several winters ago the severe frost wrought wonderful effects in my garden. The tree-branches, down to the finest twigs, appeared as if they had been turned into spun glass, and when the sun shone out the effect was beautiful beyond description. Every shrub had some special form of frost decoration, according to the shape of its leaves.

FROSTED LAUREL LEAF.

I could have spent hours in sketching the various designs, so marvellously intricate were they and beautiful, but the cold was too severe to admit of that, and I can only reproduce from memory the laurel fringes which are shown in the illustration.

The frost-needles were quite half an inch long, and

gave a curious effect to the sprays of leaves, an effect I have never seen either before or since.

Skeleton Leaves

Finding a last year's bulb turned into a skeleton by the action of rain and wind, and lying like a piece of lace-work on the surface of the ground, I picked it up this morning, and have since then been looking for such other instances of woody fibre as it may be possible to light upon in the garden and fields.

Under my holly trees were some very perfect skeleton leaves, only needing to be bleached in a weak solution of chloride of lime to form charming sprays to place with other leaves under a glass shade.

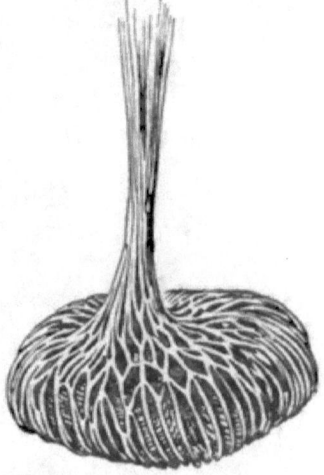

SKELETON BULB.

Magnolia leaves may often be found thus turned into skeletons when they have been lying on damp ground for some months; but as these and other specimens are seldom quite perfect, the best way, if we wish for a case of really beautiful lace-like leaves, is to make them for ourselves, by gathering well-matured specimens of suitable species, and placing them in a deep pan full of soft water, letting them soak until the upper and under skins of the leaves are rotted, when they can be brushed off with a camel-hair pencil.

When the skeletons are bleached, they should be

dried between sheets of blotting-paper, mounted into a group with fine wire and placed under a glass shade.

The following leaves succeed well :—holly, magnolia, pear, maple, poplar, and sycamore.

Seed-vessels are very beautiful when carefully cleaned.

Stramonium, henbane, poppy, winter-cherry, butcher's broom, yellow-rattle, a bunch of sycamore keys; and a very old Swedish turnip also makes a sphere of woody fibre of fine delicate network, which few people would ever guess to be the framework of that homely vegetable.

FOOTPRINTS IN SNOW

A heavy fall of snow gives us a clue to the nocturnal wanderings of such animals as hares, rabbits, foxes, rats, and mice. With a little practice, we may learn to recognise their respective footprints in the garden and fields.

Some animals run, others leap along; each creature has its own manner of getting over the ground, and what we cannot see when we catch a glimpse of them, when their limbs are in rapid action, is faithfully revealed by the snowprints.

We can soon learn hare and rabbit-marks, which always show two feet in front, one before the other, and the hind feet parallel.

The fox runs like a dog, with alternate prints, the squirrel places its short fore feet close together, and the hind feet widely apart.

Rats vary much in their movements; land and water-rats, young and old rats, all mark the snow differently, and are very puzzling to define.

February

I think mice are the cleverest little people in snowy times; they know that they can easily be seen by owls, so they form tunnels in the snow from one spot

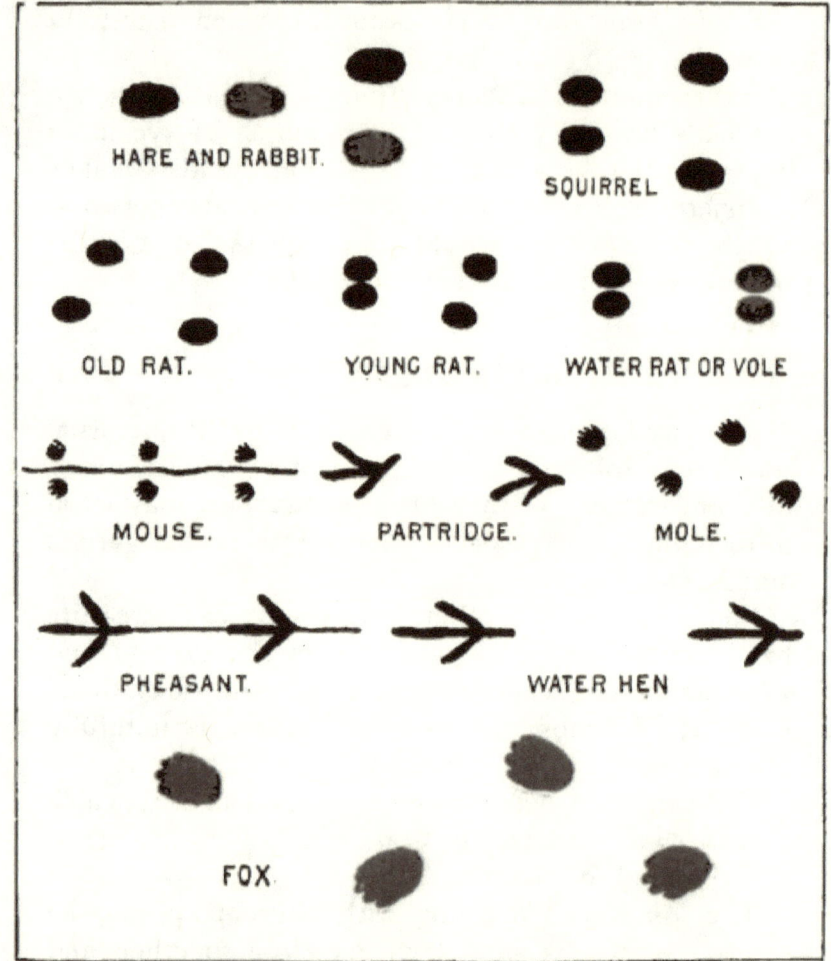

FOOTPRINTS IN SNOW.

to another, so that they may go to and fro in safety: their fore and hind feet make parallel marks as they leap along.

If a mole chances to be on the surface of the ground, he makes a furrow as he flounders through the snow, and his footprints are alternate.

I was much puzzled by a three-pronged impression always with a connecting line in the middle, but at last I discovered it to be made by the pheasant; it plants one foot exactly in front of the other, and the long hind toe makes the trailing line between the footprints.

Pigeons and doves, having very short legs, are apt to help themselves along with their wings, and these leave a sort of blurred trail rather difficult to make out, until one has seen one of these birds plodding with difficulty through the snow.

With a little study we may soon distinguish the birds that walk from those that run or hop, and once our attention has been called to this subject of footprints, we shall find it a rather amusing interest added to our winter rambles.

THE LESSER CELANDINE (*Ranunculus ficaria*)

The lesser celandine is amongst the earliest of our spring flowers.

> 'The first gilt thing
> That wears the trembling pearls of spring.'
> *Wordsworth.*

Its bright cheery flowers may be found in this month starring the ground in sheltered nooks or on hedge banks. It is one of the buttercup family, and possesses a rather curious root, consisting of small oval-shaped tubers; these break off very readily when touched, and a heavy storm of rain will sometimes wash the earth away from the root, breaking off these tubers and leaving them scattered about on the surface of the ground. This fact in olden days gave rise to

LESSER CELANDINE.

the belief that it sometimes rained wheat, as the small bulbs when detached very much resemble wheat grains.

Like all buttercups, the plant is poisonous; but in spite of that its glossy green leaves and golden flowers are welcome to our eyes, as tokens of the coming spring. It may be interesting to observe that in different flowers the petals vary in number from five to nine, and the sepals are equally varied, ranging from three to five or six.

The small honey glands at the base of the petals render this plant attractive to bees and flies, and the flowers thus become fertilised by their visits; but if, by reason of its growing in a shady place, no insects happen to visit the flowers and they fail to ripen fruit, then the plant has another resource, and produces small bulbils in the axils of the leaves, and these in time fall off and become new plants.

The name ficaria is said to be given to this species because the small tubers somewhat resemble a fig (*ficus*) in shape.

Tree Catkins

I see to-day one of the earliest signs of approaching spring! Even before the snowdrop can be found, the little hanging blossoms of the hazel, called by country children 'lambs' tails,' are to be discovered on the bare sprays. They have been there since last autumn all unobserved, but now they are daily lengthening and growing more conspicuous, and will soon be shedding out their pale yellow pollen as a passing wind shakes the branches.

From this time onward we shall find much interest in the study of tree blossoms, and I will endeavour to speak of them in the order in which they appear.

The essential thing in all flowers, in fact the very reason for the existence of a flower, is that its seed should be rendered fertile, so that when sown it should produce a plant like itself. In the greater number of plants we find stamens and pistils, which are the male

February

and female organs, contained in the same flower. In tree blossoms there is sometimes a different arrangement.

When the willow blossom is out (which we call palm), we shall find one tree bearing the pretty silvery buds which develop later on into the golden powdery blossoms; these are the male trees, and near by we

HAZEL CATKIN
(Showing male and female flowers).

shall find other willows with pale green flowers; these, after receiving a shower of pollen, will eventually bear an abundance of fluffy willow seed. Next month I shall be able to show an illustration of both these trees. In our hazel tree the female flower is at present a small brown bud, having at the apex a little bunch of crimson threads, and on the same twig hangs the male catkin with which we are so familiar. As soon as this hazel flower is fully expanded its anthers containing the pollen will split open, and the first passing breeze will scatter an abundance of the light powder into the air; some of it is sure to fall upon the crimson stigmas projecting from the brown buds; thus the future nut is fertilised, and is enabled to grow and mature into those welcome nut-clusters which we

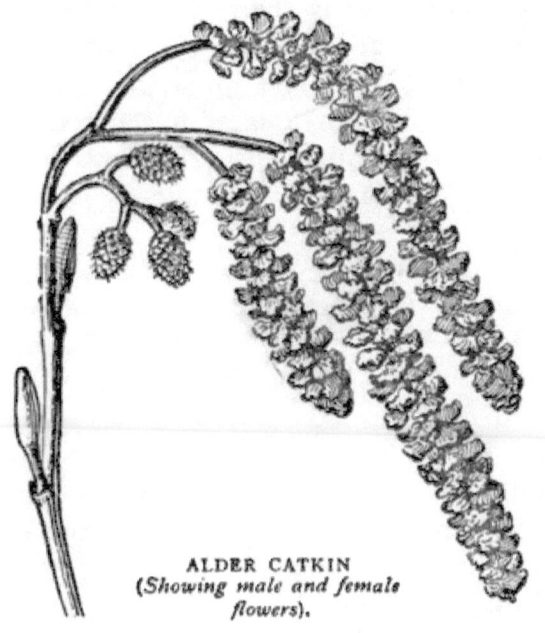

ALDER CATKIN
(*Showing male and female flowers*).

look for in the autumn hedges. Towards the end of this month we shall find the alder catkins (*Alnus glutinosa*) beginning to ripen and shed out their pollen. They somewhat resemble the hazel, only they are of a brownish red, and the future cones appear in the form of a small spray of dark crimson buds, usually found close to the hanging catkin, and it too is fertilised in the same manner as the hazel.

Queen Wasps

Queen wasps are now beginning to come out of the holes and crevices in which they have been hibernating during the winter. All the male and worker wasps die in the autumn, and only the queens survive until the following spring, when milder weather wakens them from their torpid condition, and they begin to seek a suitable place in which to build a nest and found a wasp colony.

We have in England four or five species of wasps, and each may readily be distinguished by the markings on the face and body, as shown in the illustration.

The common wasp (*Vespa vulgaris*) prefers to build either in a hollow tree or a hole in a hedge bank. Having scooped out a sufficiently large cavity, the queen lines it with a papery substance made of decayed wood.

I often watch these insects busily at work upon the stump of an old tree in my garden. With their strong mandibles they rasp off the dry wood fibres and moisten them with a glutinous liquid, secreted in their mouths, until they have a small bundle of a convenient size to carry away. With this material the wasp makes a ceiling to her nest, placing about sixteen layers one over the other, to make a firm foundation.

From this roof are suspended terraces of cells made of the same grey paper, and formed exactly like the honey-comb of bees, only these are made to contain wasp-eggs instead of honey. An egg is laid in each cell, and the grubs when hatched hang head downwards and are fed from below. This seems a curious arrangement, but the grubs are in some way enabled to hold on by their tails, so that they never fall out, and as they grow they line their cells with a kind of silk, change their skins several times, become chrysalides, and then in due time push off the cover of their cell and crawl out perfect wasps. They are pale-coloured and weakly at first,

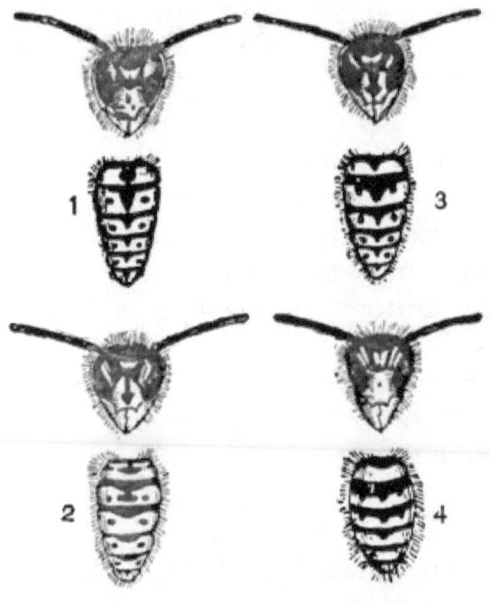

1. *Vespa Germanica.*   3. *Vespa vulgaris.*
2. *Vespa rufa.*   4. *Vespa sylvestris.*

but soon gain strength and colour and begin life on their own account.

As soon as the mother wasp finds that her eggs are beginning to hatch, she leaves off building cells, and spends her time in feeding her young brood of grubs, and goes on doing so until they are full grown. In a little while she finds herself surrounded by crowds of obedient worker wasps, and by their

February

aid she goes on enlarging the nest and laying more and more eggs, until at the end of the season a nest is said to contain as many as thirty thousand wasps.

We have reason to be grateful to these insects, because they feed upon flies, and immensely reduce their numbers during the hot summer months. I have often seen a wasp seize a housefly from the window pane and make off with it; they also pick off the teasing flies from the cattle, and thus render them valuable service.

The wood wasp (*Vespa sylvestris*) forms beautiful hanging nests in trees, where they look like grey paper roses. These nests are made of the same wood fibre masticated into extremely thin layers, forming the outer case; within are the brood cells, and at the bottom an opening is left for ingress and egress.

Although so much dreaded by most people, the wasp is really an inoffensive insect, rarely using its sting unless it is provoked and ill-treated. I cannot say as much for the honey-bee; I have known one to fly straight out of the hive and fix its sting in some innocent passer-by, who had done nothing to deserve such treatment. A bee will also pursue its victim, as I have reason to know, with unrelenting fury.

In the days when I possessed an apiary, if an ill-tempered bee set upon me, I found there was but one thing that would baffle my enemy in its pursuit; it was somewhat ignominious, it is true, to have to hide one's head in a bush and remain thus for four or five minutes, but it always proved an effectual defence; the angry hum of the bee died away in the distance, and one could at last emerge in safety.

This habit of the bee is alluded to in Deut i. 44, 'The Amorites, which dwelt in that mountain, came out against you, and chased you, as bees do.'

As we think of the life-history of the queen wasp, and how, as soon as she wakes from her winter's sleep, she sets about forming a nest, laying her eggs, and when the young are hatched feeding and watching over them with patient mother-love, and all this entirely by herself, guided only by the wonderful instinct with which she has been endowed by the Creator, I think we can but admire the qualities she possesses. And further, when we see the marvellous industry of a colony of wasps—how they also carry out the various useful purposes for which they were created, clearing away dead wood, reducing the hosts of flies, and eating many substances that would otherwise tend to pollute the air—we shall, I hope, henceforth look with different eyes upon the persecuted wasp, and instead of showing a foolish dread of its presence, learn to watch its curious ways, and recognise that it is faithfully doing, in its humble sphere, the work that has been assigned to it.

March

'In the wind of windy March
　　The catkins drop down,
　　Curly, caterpillar-like
　　Curious green and brown.'
Rossetti.

'A gleam of sunshine flashes o'er the plain,
The playful lambs amid the meadows skip,
The swallow tribe pursue their sport amain,
And in the glassy stream their pinions dip :
Though rude his greeting, all rejoice to hear
The voice of March, for well they know that Spring is near.'
H. G. Adams.

March

THE AUCUBA

AUCUBA BERRIES.

THE bright coral-red berries of the aucuba are now showing in pretty contrast with its light-green spotted leaves. This useful hardy shrub was introduced from Japan in 1783, but as it is diœcious and bears male and female flowers on different trees, no berries were ever seen on the early specimens; for it happened that they were exclusively of the female sex.

However, in 1861 Mr. R. Fortune, the great traveller and botanist, brought over from China some of the male pollen-bearing trees, and now the wind carries the fertilising dust far and wide, and the sprays of red berries appear amongst the foliage in profusion.

This shrub is not only ornamental, but has the useful

quality of thriving well in smoky air, and hence we see it frequently growing in town gardens and squares.

LAUREL-LEAF GLANDS

I do not suppose that the honey-glands of the common cherry-laurel are often observed, as they exist on the under side of the leaf, and are therefore hidden from the passer-by. We may often have wondered why, in early spring, we frequently see bees, wasps and flies buzzing about our laurel hedges, and apparently busy in collecting something which they need at that season. If we examine the back of one of the leaves, we shall discover the attraction, for at the base of the leaf and near the midrib are from two to four glands exuding a sweet liquid which affords welcome sustenance to insects. What particular use these glands may be to the shrub itself is not known; they seem to be a speciality of the laurel; for, although I have examined a large number of shrubs and trees, I cannot find similar glands in any other plant, though doubtless some may exist.

The so-called laurel is really a species of cherry, and in favourable years it bears long sprays of purple berries. The true laurel is the bay tree, *Laurus nobilis;* it also bears cherry-like fruits, but only in the southern parts of England.

THE MEALWORM BEETLE (*Tenebris molitor*)

When our feathered pets are of a kind that will not prosper without insect diet, the best mode of supplying them, during both winter and summer, with food which will keep them in health and vigour, is always rather a difficult problem.

Ants' eggs are collected and dried, and can always be purchased throughout the year, and these afford a useful food for many species of birds, although I have not found them always approved of by my own special pets.

Raw meat is another resource, but it is troublesome to prepare, and very difficult to keep fresh in hot weather. The one item that seems indispensable in bird-keeping is the mealworm, and, as many people have asked me, 'What is a mealworm?' I will take it for my subject to-day. If my readers will refer to

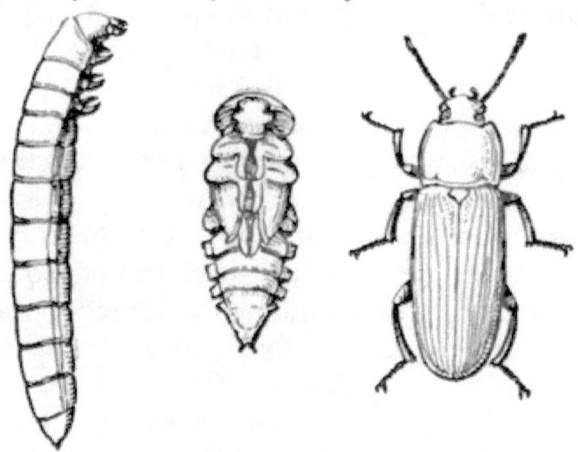

MEALWORM BEETLE.
Magnified about six times larger than life.

the illustration, they will see a long sort of caterpillar, which is the aforesaid mealworm, the larva of the mealworm beetle.

Instead of being soft, like an ordinary grub, it is hard and polished, of a brownish yellow colour, and in all respects extremely like the destructive pest called by gardeners the wireworm. The latter, however, is the larva of a different species of beetle which feeds on plant roots.

The mealworm beetle is always to be found in

mills, granaries, and bakehouses—in fact, wherever flour is kept, for in it the beetles lay their eggs, and these hatch into minute thread-like grubs, which in two years' time grow into flat long-bodied mealworms, perfectly harmless, scentless creatures, easily kept in a tin box filled with barley-meal and flour. They grow and fatten all the quicker if the box is kept in a warm place and some layers of flannel are supplied, as they feed upon flannel as well as upon flour. The flannel should be moistened occasionally with a little beer or water.

At length the worm turns into a curious mummy-like chrysalid, and then into the perfect beetle, which, although it is black, is not in any way related to the so-called blackbeetle or cockroach (which is not a beetle at all), and is of a reddish brown, the male possessing four strong wings.

The mealworm beetle is as innocent and harmless as its larva; I sometimes find a wandering specimen near my bird-cages, and I know I can safely pick it up and restore it to the box where its kith and kin reside, with no fear of its biting or leaving any odour on my hands.

The English nightingale is unfortunately so greedy for this insect, that birdcatchers can always trap it with the greatest ease by clearing a space upon the ground and placing some mealworms on limed twigs. The bird flies down immediately to secure the dainty, and is held fast and caught by the snare so cunningly set.

WHITLOW GRASS (*Draba verna*)

Early in this month I found, on an old wall, the pretty rosettes of one of our very early spring flowers, the whitlow grass, which is not really a grass, but a miniature plant, seldom more than three inches high, and sometimes so small that it only occupies a space that might be covered by a shilling.

WHITLOW GRASS.   RUE-LEAVED SAXIFRAGE.

On a tiny central stalk it bears a few white flowers, which droop gracefully when the air is moist; the petals quickly fall away, and then small oval seed vessels appear; these, when mature, shed off two outer husks, leaving a white membrane which divides the seed vessel, just as one sees it in the seed vessel of the common honesty.

During February and March the whitlow grass may sometimes be seen growing in such profusion on old ruined walls as to give the effect of a slight fall of snow.

Another charming little annual which haunts old walls is the rue-leaved saxifrage (*Saxifraga tridactylites*). It rarely exceeds three inches in height; a dainty little plant with white flowers, three-lobed leaves thickly covered with viscid hairs, upon which small insects may often be found entangled.

When the flowers are over, the stem and leaves become of a rich red tint, which seems frequently to be the case with plants exposed to the full sunlight, as they are when growing upon rocks or walls. We may prove this by trying the experiment of keeping two specimens of this plant in pots, and placing one of them in a sunny spot and the other in shade. We shall find that the latter will continue to be green, and fail to attain its natural crimson colour.

THE DANDELION

Dandelion flowers are now making such a bright glow of colour by the roadside that we will choose them for our subject of study to-day.

The plant takes its name of dent-de-lion from the form of the leaves, which are so deeply cut as to resemble teeth; more especially, perhaps, in the spring is this the case, as later on in the summer they become less sharply indented.

The flower-bud rises from the centre of the plant to nearly a foot in height, then it opens and becomes fertilised by insects. As soon as this process has been completed, the flower closes up, and the dead petals and calyx leaves remain like a pointed roof defending the seed from rain. Now the stalk bends down until it lies flat upon the ground, where it remains about twelve days. By that time the seeds are matured, and the stalk again rises to an upright position. The calyx leaves now turn back until they are parallel with the stem, and the beautiful downy globe is formed and expands until it is a fluffy ball of seeds, hanging so loosely that the lightest breeze can waft them into the air.

The seed itself is worth examination. When, after a longer or shorter flight, a seed touches the ground and falls into some crevice, it might still be dragged out by the wind and carried away; but this is guarded against by some spiny projections on the upper part of the seed, which tend to hold it securely in its place.

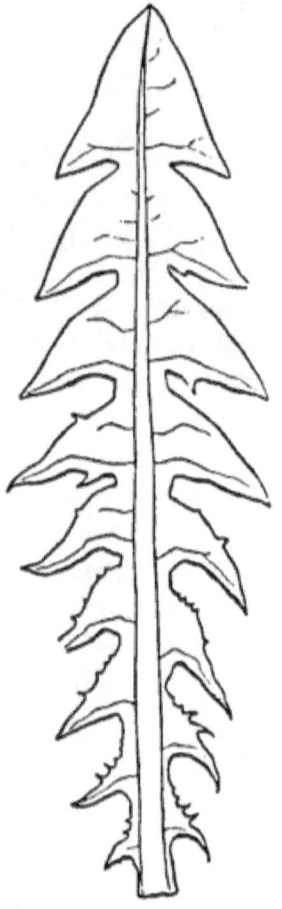

DANDELION LEAF.

The Nuthatch (*Sitta Europæa*)

The nuts we throw out at the windows for the squirrels are frequently shared by the nuthatches. These pretty birds abound in my old garden, and in the course of years they have become so extremely tame that they will almost take nuts out of our hands.

An old oak tree on the lawn near by is much used by these birds; they ram the barcelonas into crevices in the rugged bark, and whilst they hang head downward to gain the greater force, I hear the beaks' loud hammering going on, and afterwards find the empty nutshells, from which the kernels

NUTHATCH AND NEST.

have been extracted, still remaining in the interstices of the tree bark.

The loud call-note of this bird is one of the early signs of coming spring. It is hard to believe that the small feathered creature that we see creeping up a tree-stem like a grey mouse can be filling the woods with so much sound. Its mating call-note is a clear sharp cry, several times repeated at short intervals, and maintained throughout the early spring months.

One ancient lime tree near this house has frequently been the nesting home of four species of birds. In the highest hole some starlings established themselves. Just below a smaller cavity was taken by a pair of nuthatches. Some jackdaws appropriated another opening in the stem, and lower down a neat round hole was bored by a green woodpecker.

These various lodgers all appeared to live harmoniously together, and they allowed me to watch them as they flitted in and out on family cares intent. The green woodpecker was the most wary, and would seldom allow me more than a hasty glimpse of his crimson head and golden green plumage.

The nuthatch has a curious habit of closing the entrance to its nest with layers of mud until only a very small hole remains. The illustration shows a case in point. The bird had made its nest about twelve inches down a hollow tree trunk, and then, with infinite labour, it brought yellow clay sufficient to close up the tree stem, leaving but a small hole for ingress and egress.

It is said that the male bird keeps its mate upon the nest, and feeds her through the entrance hole until her eggs are hatched. I have not seen this myself, and can only give the fact as stated by others.

Tree Seeds

The high winds which usually prevail in early spring are performing a very useful office in scattering the seeds of trees and plants. When the hornbeams, sycamores, and maples have been unusually full of fruit, their dried bracts and seeds will be found lying thickly strewn over the lawn.

The winged part of the sycamore fruit (botanically

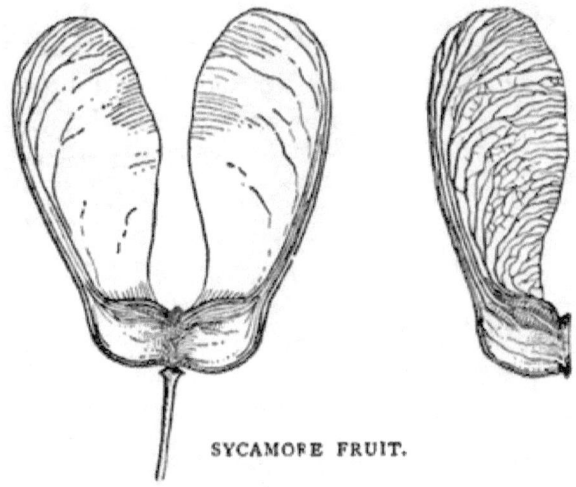

SYCAMORE FRUIT.

called *samara*) has in many cases become a delicate piece of lace-work, the action of rain and wind having made it into a skeleton; the heavy end is entangled in the grass, and out of the seed-case a

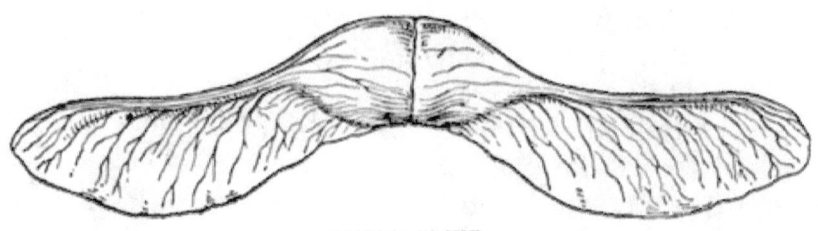

MAPLE FRUIT.

young rootlet is finding its way into the ground. Later on, I shall be able to find and record the unfolding cotyledon leaves, which are curiously rolled up within the seed-case.

The maple fruit is also two-seeded, and somewhat

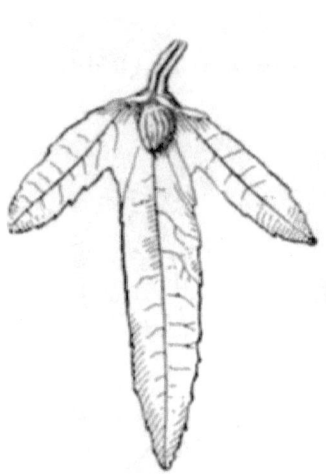
HORNBEAM SAMARA.

HORNBEAM SAMARAS RIPE.

resembles the sycamore (which is a maple), except that the two *samaras* are joined at a different angle. The fruit quickly divides, and each seed has then a fair chance of germinating.

The hornbeam *samara* is like a three-pointed leaf, the sharp-angled nutlet being attached to it at the lower end. Each bract in the cluster seems to prepare for its flight by breaking off from the stem, and then hanging by a hair-like thread, so that a passing breeze may easily detach it.

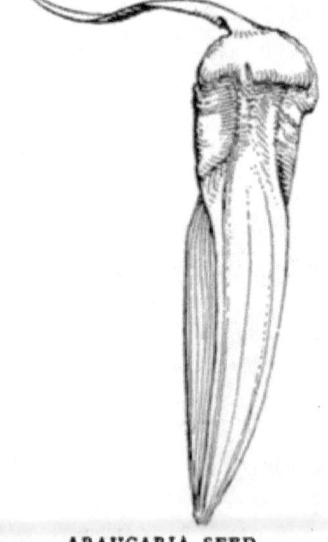
ARAUCARIA SEED.

The araucaria, when well established and growing vigorously, will sometimes produce its huge cones in England. There are male and female trees; the cones of the former will shed out more than a wineglassful of yellow pollen. The strange-looking seeds which fall out of the fertile cones are sure to attract attention as they lie on the grass by their peculiar form and large size.

POPLAR CATKINS

The flowers of various species of poplar are now appearing, and form an interesting subject for study. I have obtained to-day the catkins of the aspen (*Populus tremula*), the abele or white poplar (*P. alba*), the Lombardy poplar (*P. nigra*), and the grey poplar (*P. canescens*). A slight shower had brought out the perfume of the buds and blossoms of the balsam poplar or tacamahac (*P. balsamifera*), which has very conspicuous catkins of a bright reddish-brown.

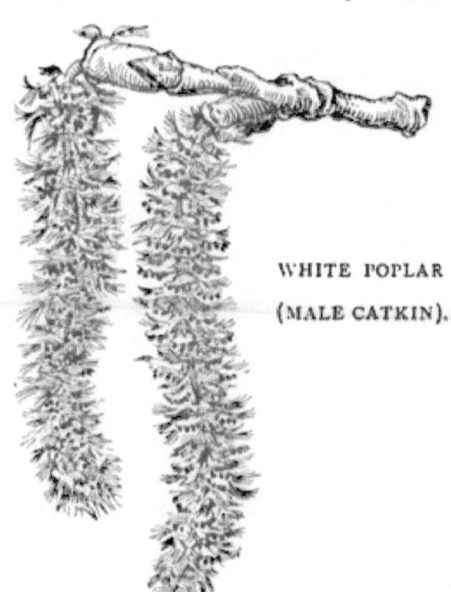
WHITE POPLAR (MALE CATKIN).

As most of these trees flower mainly on the upper branches, where we cannot reach the catkins, we must be content to pick them up, as I did to-day, beneath the trees, where they look extremely like red and brown caterpillars.

Poplars are all diœcious trees; that is, bearing flowers with stamens on one tree and flowers containing pistils on another, usually growing near by. This makes their study rather puzzling, and it is further complicated because the willows are now in flower, and there is a certain resemblance between them. We may, however, always recognise poplars by their drooping catkins, whilst willow flowers are invariably borne upright upon their stems. The male catkins bearing the stamens are usually the most conspicuous, and often they appear earlier than the female flowers.

By dissecting a specimen poplar catkin from each tree, we can readily trace the different parts, the fringed scales bearing the stamens and small woolly stigmas which catch the pollen-dust brought them by the wind.

WHITE POPLAR (FEMALE CATKIN).

Poplar catkins are usually fertilised by the wind; they contain no honey, and are therefore unattractive to insects. The willows, having small honey glands, offer three lures to the insect tribe—colour, scent, and honey—hence we may be sure to find bees and flies frequenting their early blossoms.

MALE FLOWERS OF YEW.

THE YEW TREE

'When the rude natives of this polished land
Formed the strong phalanx of their valiant band,
With dext'rous hand the bended bow they drew,
And shaped their arrows from the dusky yew.'
                                FRANCES A. ROWDEN.

The male blossoms of the yew tree are now fully out, and as a passing breeze shook the branches to-day they sent out clouds of yellow pollen. I never see this happening without recalling Tennyson's interesting allusion to the 'smoking' of the yew tree.

'O brother, I have seen this yew tree smoke,
Spring after spring, for half a hundred years.'
                                The Holy Grail.

The female blossom is on a separate tree, and it may be at some distance away; the wind therefore carries the fertilising pollen far and wide, and in due time it reaches the other flower, which will eventually produce the beautiful wax-like yew berries.

I used to think that the showers of pollen, which make the ground under the tree look yellow with its abundance, was an instance of needless waste; but I have now observed that many species of flies and solitary bees are ex-

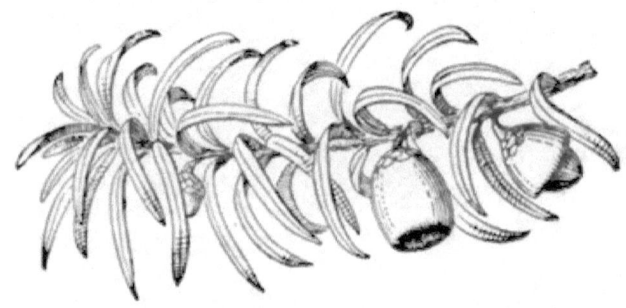

YEW BERRIES.

tremely fond of pollen and feed greedily upon it, as well as use it to store in their nests for their young grubs to feed upon when hatched. Doubtless in this way the tree is able, all through the early spring, to afford the winged creatures an abundant supply of needful food until they are able to obtain honey from the summer flowers.

WILLOW CATKINS

As one of the tokens of coming spring, it always gives me a thrill of pleasure to note for the first time the silvery willow buds appearing. As the dark brown bud scales begin to open and reveal the silky down within, then, as the sun gains power, these outer scales fall off, and the pure white catkins become conspicuous. They daily grow in size, until, attaining maturity, they are covered with pollen of a rich golden yellow. This

MALE. FEMALE.
SALLOW CATKINS.

pollen is highly attractive to the newly awakened humble bees. These may be seen clustered upon the blossoms, not only feeding themselves, but carrying away provision with which to store their cells.

It is interesting to observe that while the willow has only one bud scale, the lime tree has two, and other trees usually have many outer coverings for the bud.

The male and female catkins are shown in the illustration, and, as I have said, they grow on different trees, which are usually found within a short distance of each other, so that the wind may carry the pollen from one tree to another in order to fertilise the flowers.

A small low-growing species of willow called sallow, which, by the way, grows abundantly on our common, is the kind which is most frequently gathered for decoration at Eastertide. This custom dates from the time when palm-branches were strewn before our Lord when He was riding into Jerusalem.

The true palm, of course, is still used in Eastern countries for church decoration; but as we in England have no tree with either fresh green leaves or conspicuous blossoms flowering at Easter, the willow, with its pretty golden catkins, has been called palm, and substituted for it for many generations. A passage of Goethe on this subject has been thus translated:—

> 'In Rome, upon Palm Sunday,
>   They bear true palms;
> The cardinals bow reverently,
>   And sing old Psalms.
> Elsewhere their Psalms are sung
>   Mid olive branches;
> The holly-bough supplies their places
>   Among the avalanches;
> More northern climes must be content
>   With the sad willow.'

With reference to that last line it is rather curious that, from the days when captive Israel hung their harps upon the willows of Babylon, the tree should have been regarded as an emblem of sadness: and yet, in later times, it should have changed its character, and become a token of joy and gladness.

We possess from thirty to forty kinds of willow in Britain, ranging from trees eighty feet in height down to the dwarf species which abound on northern moors, and are only a few inches high.

I have gathered sufficient of the white silky down from the willow seed-vessels on our common to stuff a sofa-cushion, and in fine weather the air is filled with the light fluffy seeds which are thus carried far and wide.

We owe to the willow the valuable medicine salicine, so much used for the alleviation of rheumatic pains. A preparation of salicine crystals forms a beautiful microscopic slide, and when shown with the polariscope exhibits exquisite rainbow colours.

April

'Young leaves clothe early hedgerow trees;
Seeds and roots, and stones of fruits,
Swollen with sap, put forth their shoots;
Curled-headed ferns sprout in the lane;
Birds sing and pair again.'

Christina Rossetti

April

St. Mark's Fly (*Bibio Marci*)

ST. MARK'S fly, so-called because it generally appears about the time of the saint's day, has come late this year; but I see it now resting on various flowers, or else flying in its own very peculiar way, with its long hairy legs hanging down like a bunch of black threads.

The male fly has clear wings, those of the female are dusky; the former has eyes double the size of those of the latter; both the insects are jet black, and very sluggish in their movements, so by these characteristics they may be easily identified.

The female lays about one hundred and fifty eggs at a time in grass roots or decayed vegetable matter, upon which the grubs feed. These remain in the ground throughout the winter, and when full grown the larvæ become chrysalides, and in a few weeks' time the perfect flies emerge.

Another fly belonging to the same genus (*Bibio Johannis*) is called St. John's fly, as it is to be seen about the latter end of June, when St. John the Baptist's day falls in the calendar. I am not familiar with its appearance, but I imagine from its scientific description it must be very similar to the St. Mark's fly.

These two insects are, I believe, quite harmless,

but some of their near relatives are grievous torments to horses and cattle in the various countries where they are found.

In Servia a minute fly so irritates the flocks and herds by its intolerable stings, that hundreds of sheep and oxen are driven mad and perish in consequence of its attacks.

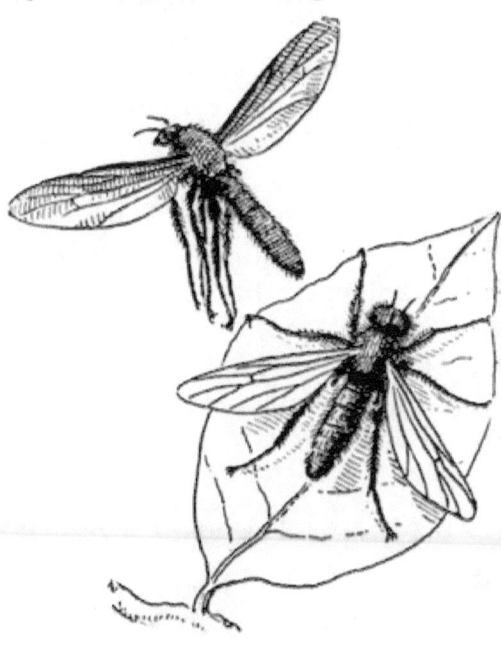

ST. MARK'S FLY.

In India there are flies that can even pierce the elephant's hide, and in Florida, cows, horses, and mules are almost eaten alive by voracious fat-bodied flies, which give them no peace during the summer months. It is rather a consolation to know that an insect called the 'coachman fly' preys in its turn upon these tormentors,[1] and 'will sit through a long drive on the collar or some other part of the harness, or even on the steed itself, in order to pounce upon the insects as they settle. The curious thing is that the horses seem to know the difference, for directly a horse-fly comes, even if it does not sting, they become restless, tossing their heads and lashing with their tails; but the "coachman" may rest on any part of

[1] *Royal Nat. History*, vol. vi., p. 59.

them for any length of time and never be interfered with or driven off.'

The tsetse fly of Africa is perhaps the most formidable of these insect plagues; its bite is fatal to horses, oxen, and dogs. Dr. Livingstone was constantly hindered in his missionary journeys by this apparently insignificant enemy, for in one short journey, although he scarcely saw more than twenty of the flies, yet forty-three of his valuable draught oxen died from their attacks.

The tsetse fly is scarcely so large as a bluebottle, of a brown colour, with yellow markings, and a long proboscis; fortunately, its bite is harmless to man, but travellers may well dread its peculiar buzz, as it may portend the death of their horses and cattle, by means of which alone they can journey across the African deserts.

THE DEATH'S HEAD MOTH (*Acherontia atropos*)

I had a surprise this morning. A splendid specimen of the death's head moth (*Acherontia atropos*) came out of its chrysalis, and was reposing upon a small branch I had placed for its convenience. For seven months I have been tending this said chrysalis, keeping the moss on which it rested sufficiently damp and yet not too wet, as either extreme would have been fatal to the insect.

Never having seen a living specimen of this, one of the largest of our native sphinxes, I gazed with delight at the varied markings on the body and wings, a rich intermingling of brown, blue, fawn and velvety black.

The antennæ are black and end in a white hooked

bristle. The legs are barred with black and white, and thickly clothed with fawn-coloured masses of furry down. With bright orange under wings and a portly body of pale blue and orange, my readers can believe my new acquisition is indeed a rich piece of colouring.

DEATH'S HEAD MOTH AND LARVA.

The singular mark upon the thorax from which the moth derives its name indistinctly resembles a human skull ; an unfortunate fact for the insect itself, as in olden days it was looked upon as a weird forerunner of all kinds of evil, and its also possessing the power of emitting a low squeaking sound was sufficient to raise up a host of superstitious fears in the minds of ignorant people who persecuted and

April

killed it without mercy. The Rev. J. G. Wood[1] relates an amusing incident where 'A whole circle of village people were standing around a death's head moth that had by some mischance got into the churchyard. Not one of them dared to touch it, and at last it was killed by the village blacksmith, who courageously took a long jump and came down on the unfortunate moth with his iron-shod boots.'

I hoped to feed and tame this curious sphinx, but it will not partake of any kind of food, not even honey, which is said to be so attractive to this species of moth as to lead it to force its way into beehives, much to the annoyance of the bees; they are sometimes compelled to raise waxen walls at the entrance of the hive to keep out these intruding moths.

The caterpillar of this sphinx varies much in colour, but is usually of a lemon yellow and green, with violet stripes on its sides; it is often four or even five inches in length. It feeds on the potato, jessamine, and deadly nightshade, but is not often found, because it hides itself in the earth during the day, and creeps out for its food at night. When labourers are digging up potatoes they frequently find the great chrysalides of this moth, which they invariably call 'locusts,' 'ground-grubs,' or 'maggots.'

I obtained my specimen from a poor woman who was begging her way to some potato-fields where she hoped to obtain work. I learned that she often came across these 'locusts,' as she called them, when engaged in digging up potatoes, and having received an order for some she duly brought them to me, but unfortunately only one chrysalid has survived the winter and reached the perfect stage.

[1] *Insects at Home.*

The Humble-Bee Fly (*Bombylius major*)

The appearance of the graceful humble-bee fly hovering over the early spring flowers is to me one of the welcome signs of spring.

It flutters over my beds of forget-me-not and pulmonaria, and poising on the wing like a humming-bird, it inserts its long and very slender proboscis into each blossom in succession, extracting the honey upon which its delicate life is sustained. The slightest movement on my part sends it off so swiftly that the eye cannot follow it, and yet it will return after a time, and allow me to watch its graceful flights just as long as I remain perfectly still.

It is a fly with a good deal of character, and it differs in many respects from any other with which I am acquainted. I have sometimes caught a specimen in a soft gauze net, and carefully placed it under a glass shade containing a small vase of sweet flowers for its refreshment. At first the fly gives up all for lost and lies on its back with its slender legs in the air, as if in a dead faint; but it soon revives, and softly humming to itself, it hovers gently round the flowers, and when at last assured that there is no outlet for escape, it becomes quite resigned and begins to draw honey from the blossoms until it is satisfied, when it will rest upon a leaf in a contented fashion, not in the least minding its loss of liberty.

If my readers will contrast with this the conduct of a newly caught bluebottle fly placed under a glass, and think of the wild way in which it will strike itself against its prison walls, buzzing and dashing about in a blind unreasoning fright, I think they will understand what I mean by difference of character in insects. This might afford a very interesting subject for study.

I believe very little is known about the life-history of this charming insect. Its larvæ are said to be parasites, feeding upon caterpillars and other insects. The perfect fly is seen from March to May, but I have not observed it in the summer or autumn months.

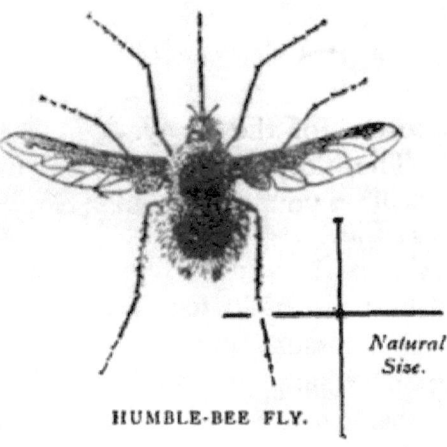

HUMBLE-BEE FLY. *Natural Size.*

THE ASH

The ash is now becoming conspicuous by the size of its dark flower and leaf-buds. This feature has often been noticed by the poets; Bishop Mant speaks of

> 'Its buds on either side opposed
> In couples each to each, enclosed
> In caskets black and hard as jet,
> The ash-tree's graceful branch beset.'

I scarcely ever pass by an ash tree in spring but I recall Tennyson's well-known lines—

> 'Those eyes
> Darker than darkest pansies, and that hair
> More black than ashbuds in the front of March.'

As a rule the buds are placed exactly opposite to each other on the branch, but in the illustration they are alternate, as I find is often the case towards the end of the spray.

The flowers of the ash are varied enough to puzzle a young botanist. Some of the flowers contain

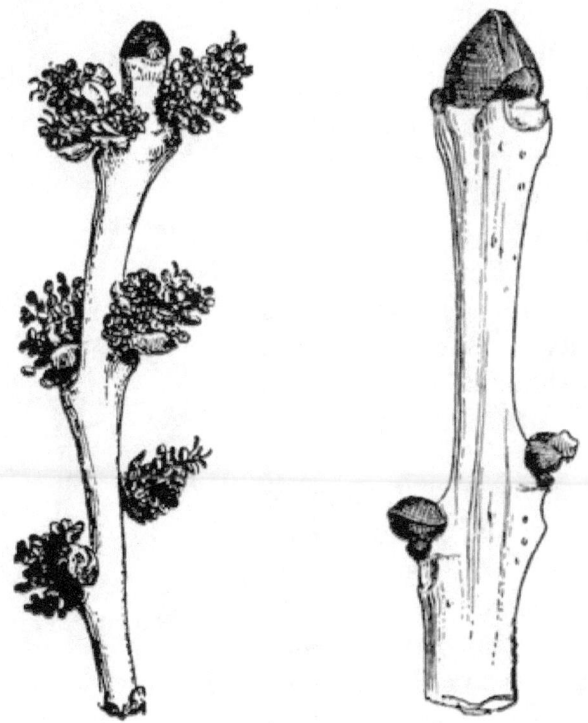

FLOWERS OF ASH.

stamens, others bear only pistils, and some may be found with both stamens and pistils; these varied blossoms are described in botany as polygamous.

The ash is largely grown in Kent to supply poles for the hop-grounds. The trees are planted in narrow strips of ground adjoining the fields, and when the young saplings are sufficiently tall they are cut down, and after a few years the stems that have sprouted from the root-stocks are just the straight poles that are required to support the hop-plants. The process is repeated from time to time, so as to maintain the needed supply. A little wood of this kind is called a 'shave,' possibly a corruption of the word shaw, with which we are familiar.

In olden times the ash was called 'The Husbandman's Tree,' as it supplies tough, flexible handles for all kinds of tools and agricultural implements.

We may easily distinguish the two kinds of catkins on the birch; the pistil-bearing flower is small and upright, whilst the male catkin hangs down and bears the pollen in its bracts. Towards the autumn we shall find the small catkin, which is now erect, will have become pendent and composed entirely of minute seeds which autumnal gales will carry far and wide.

BIRCH CATKIN.

SEEDLING TREES

The lawns and flower-beds are now covered with sycamore, beech, and other seedling trees in various stages of growth. As the two seed-leaves, or cotyledons, as they ought to be called, differ very much from the mature leaves, it is rather interesting to try and find out each species, and thus learn to identify trees in their babyhood.

The sycamores seem to find it difficult to get out of their seed-coats, for here and there we may find one with a stem an inch long with the winged part (*samara*) perched at the top like some quaint kind of headgear. Even if they get out of the husk, they are for the first day or two crumpled into odd shapes, just as they were packed and curled up in the seed-coat; but before long they spread out their two cotyledons and seem to rejoice in the light and air.

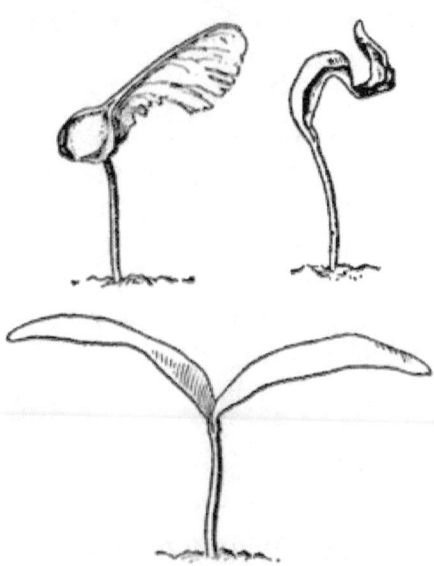

SYCAMORE SEEDLINGS.

These seedlings are of a dark green colour with a crimson stem, a combination we may also find in the bud of the tree itself, which in some specimens has outer bud-scales of the richest crimson, whilst the delicately folded young leaves within are of a vivid tint of green.

I have just found a remarkable number of these seedlings with three and even four cotyledon leaves, showing that the seeds must have contained three or four embryos A month or two later these young trees will show two leaves of the mature form, which is quite different from the strap-shaped cotyledon.

It may interest my readers to be reminded that the sycamore of which I have been speaking is not the sycamore of Scripture, which is a species of fig, and an entirely different tree in every respect. It has an oval undivided leaf like the bay tree and having wide-spreading branches affording abundant shade; it is often to be met with by the roadside in Palestine, where it is planted for the benefit of wayfarers, who welcome the cool shelter it affords from the hot sun, and also the sustaining fruit it bears.

Beech seed-leaves consist of two broad, deep green cotyledons of palest green beneath; they are very distinctive, and once identified we can never mistake them.

The lime tree has seedlings with deeply incised leaves, very unlike the perfect form.

If we have no companion who can name these baby-trees for us, the only way to learn about them is to look under the various trees in April and May, when we shall probably light upon the growing seeds of each kind. When pressed and dried, they form an interesting collection either by themselves, or to add to any dried specimens of the English forest trees we may happen to be forming.

SHEPHERD'S PURSE

One of the commonest weeds to be found throughout spring and winter is the shepherd's purse (*Capsella-bursa pastoris*). It often bears as many as fifty pods on its stem, and by counting the number of seeds in each pod and adding the whole number together, we shall find the total to amount to about one thousand five hundred seeds. No wonder gardeners find it a troublesome weed, when one plant can produce so many seeds and sow itself all over the garden. We may note its very varied leaves, those on the stem are oblong and arrow-shaped at the base, the root leaves being pinnatifid, that is, cleft into divisions half way down.

In China and North America the plant is used as a vegetable, and it used to be credited with medicinal virtues.

My chief interest in this hardy little weed arises from its remarkable power of adaptation; if it happens to be growing in rich soil, it will attain to a height of one or two feet, but if starved in some wall crevice or growing between the stones on a hard gravel path even there it does its best; its dwarf stem is covered with immature purses, and is crowned by a tiny head of flowers; it is thus a true emblem of patience and fruitfulness under adverse circumstances.

SHEPHERD'S PURSE.

May

'All the land in flowery squares,
Beneath a broad and equal-blowing wind,
Smelt of the coming summer, as one large cloud
Drew downward; but all else of heaven was pure
Up to the sun, and May from verge to verge,
And May with me from head to heel. . . .
                    To left and right
The cuckoo told his name to all the hills;
The mellow ouzel fluted in the elm;
The red-cap whistled; and the nightingale
Sang loud, as tho' he were the bird of day.'

Tennyson.

May

Honey Guides

RHODODENDRON.

IT is interesting to observe the markings upon the petals of flowers which serve as honey guides for the bees. For instance, in the rhododendron the stamens all curl upwards, and the richly coloured spots are placed on the upper petals, to direct the bee where to alight. As it passes down into the flower to obtain the honey it is seeking, it cannot help brushing pollen off the anthers, and thus, its hairy body becoming covered with the powder, it carries it to the next flower it enters, and ensures what is called cross fertilisation—

that is, the pollen of one blossom being placed on the stigma of another.

In the gladiolus the stamens are differently arranged, and the bee is required to enter below instead of above the stamens; there are therefore three honey guides on the lower petals, and the bee, all unconsciously, bears a load of pollen on its back, and performs its useful office of fertiliser to each flower in succession.

In the iris the lower petal is usually covered with a rich pattern of coloured stripes, which all lead up to the narrow passage where the bee must enter and push its way, necessarily brushing pollen off the anther in its progress to reach the honey at the base of the petal; as it enters the next flower, it cannot fail to leave the pollen on the stigma at the entrance; and this wonderful contrivance can be traced in the delicate stripes of the wood-sorrel and very many other flowers, where distinct way-marks are afforded to guide the bees in their most useful work of fertilisation. It adds an interest to our walks to know that the infinitely

GLADIOLUS.

varied beauty of flower-tints and markings have this useful purpose in view.

The close connection that exists between insects and flowers has been much studied of late, and it has been ascertained that many plants cannot produce seed unless their flowers are visited by insects. When orchard-houses were first built, and stocked with peach, nectarine, and other trees, scarcely any fruit was produced, because no provision had been made for allowing bees to enter and do their useful work.

This was the case in my own peach-house years ago, so a bee-hive was introduced when the blossoms were ready for fertilisation, the busy insects did their work effectually, and a good crop of fruit was the result; but the poor bees could not find their way back to the hive, and they nearly all died. To obviate this sad disaster, the gardener has learned to fertilise peach-flowers by brushing them lightly with a hare's foot, which detaches the pollen and conveys it to the anthers without injuring the blossom.

The Common House Fly (*Musca domestica*)

It is rather surprising that, as a rule, so little is known about the life-history of the common house fly. The creatures abound in our houses, we have been familiar with them from childhood, but where they come from, how they propagate, and what are the stages of their life-history, who can tell us?

Perhaps it may be interesting to throw some light upon this domestic plague, and more especially will this be useful because a little knowledge about flies will enable us greatly to reduce their numbers.

The common house fly lays its eggs in vegetable

refuse, decaying cabbage stalks and such like; it is therefore important that such matters should be burnt instead of being thrown into the dust-bin. The eggs hatch into small white grubs: these, when full grown, become chrysalides, and the flies emerge in due time.

The bluebottle fly is only attracted by a meat diet. These flies find out any dead animal or bird, and quickly deposit dozens of very small white eggs upon it. The eggs hatch out in a few hours into small white maggots (known to fishermen as 'gentles') of a peculiar shape, being pointed at one end and flat at the other.

LARDER FLY (*Magnified*).   BLUEBOTTLE (*Magnified*).

These creatures devour any kind of flesh with wonderful rapidity, so that Linnæus declared that 'Three bluebottles could eat an ox as fast as a lion could.' The bluebottle is a very determined character. Even when meat is covered by a wire sieve this insect will often drop its eggs upon the joint through the interstices of the wirework, so that to make a larder really fly-proof the windows should be protected by fine wire gauze. The smaller greenbottle fly has the same habits, and spends its life in laying eggs on dead or decaying substances or else basking on leaves in the sunshine. When seen through a magnifying glass its body glistens like a precious stone, or like burnished golden-green metal; although this insect is so common it is well worth examination, for its beauty really baffles description. Its many faceted eyes and the formation of its feet should also be observed by the student.

The larder fly (*Sarcophaga carnaria*) is the largest of the genus, being half an inch in length; it differs from the other flies in depositing its young alive upon decaying animal and vegetable matter, and, sad to say, it sometimes places its grubs upon living animals.

Réaumur calculated the number of young produced by one fly of this species to be about 20,000; we may therefore imagine how valuable such an insect is in speedily removing decaying substances which would otherwise tend to pollute the air.

CANARY GRASS (*Phalaris Canariensis*)

Although bird-keepers are familiar with the canary-seed with which they feed so many of their pets, yet comparatively few people see the canary-grass growing, or even know that there is such a grass.

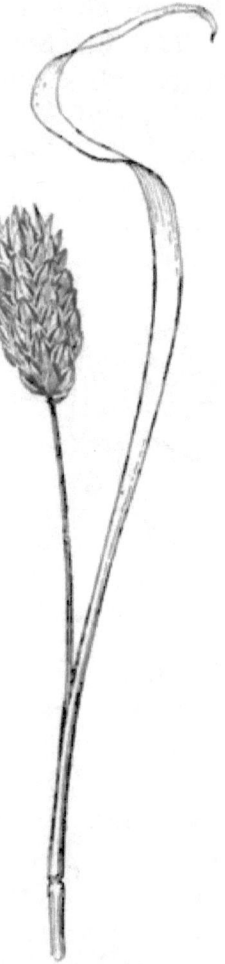

CANARY GRASS.

I am apt to have a patch of it sown in some of the garden beds every year, as it is a beautiful sea green colour and makes a charming variety with other flowers. The stems are about two feet high, the leaves lance-shaped, and the soft round heads of flower are pale green streaked with darker markings.

If we have but a few pots on a window-ledge, canary grass can be grown. About a dozen seeds sown in good soil and kept watered and sheltered from frost, will result in our seeing the pretty flower heads in due time. April or May would be the best time to sow the seed either in a pot or in the ground.

Canary grass is said to have been cultivated in this country in order to supply singing-birds with food ever since the days of Queen Elizabeth. It was introduced from Central Asia. It is largely grown in Kent and in the Isle of Thanet.

THE TRINITY FLOWER (*Trillium erectum*)

This plant rejoices in a variety of names. In North America it is known as the wood-lily, three-leaved nightshade, and Indian shamrock; its Latin name is *Trillium*, the number three seeming to be the order of its being. It possesses three leaves, three green bracts, which look very much like the sepals of a calyx, and three perianth leaves, differing from petals only in that those terms petal and sepal are never used in describing plants of the lily family.

I watch for the flowering of my *Trilliums* each spring with keen interest, not only for their own exquisite beauty, but also on account of the halo of poetic charm woven around this flower by Mrs. Ewing in her sweet legend of *The Trinity Flower*. I will not attempt to quote from it, but would advise my readers to obtain the little book[1] in which it may be found, and then they will be able to understand my reverent love for this charming flower. My plants were imported some years ago

THE TRINITY FLOWER.

from Massachusetts; but they now can easily be obtained from dealers in herbaceous plants at home.

Trillium grandiflorum has large snow-white flowers, and is the most beautiful of the sixteen species.

[1] *Dandelion Clocks*, by Mrs. J. H. Ewing.

The illustration is drawn from *Trillium erectum*, which is called in America beth-root, Indian balm, and lamb's quarters. It has green bracts striped with purple and reddish-purple perianth leaves. From its root a medicine is prepared which is valued for its curative properties.

This wood-lily is perfectly hardy, only requiring a light soil and a shady damp situation. It comes up year after year, appearing in April and flowering early in May.

Flowering Trees

In this month so many different trees produce their flowers or catkins that we must be on the alert to study them before they fall to the ground or are blown away by the wind.

Nature keeps us breathless in the attempt to overtake her marvellous energy. Every day something fresh appears; wild flowers are springing up, buds are opening, even early horse-chestnuts are to be met with in full leaf, and growth is so rapid under the increasing warmth of the sun that sprays of opening buds which we may be wishing to paint are expanded into leaves before we have time to record their beauty in an early stage. Amongst other trees we must not fail to notice the hanging sprays of larch with their yellow stamen-bearing flowers, the pollen from which falling upon the delicate crimson blossoms on the same spray will enable them to become the cones of next autumn, the wind being the agent in this process.

Many years ago, before the cultivation of the larch was understood, two seedling plants were sent to the

Duke of Athole; and his gardener, with the best intentions, treated them as hothouse plants, which speedily brought them to such a dying condition that they were thrown away upon a rubbish heap. Hardly had they taken up this ignominious station when they revived and began to grow. When I visited Dunkeld many years ago the guide pointed out with pride two magnificent larches, which were the aforesaid specimens now flourishing under favourable conditions.

The larch is a native of the Alps, and the roofs of the picturesque *châlets* in Switzerland are covered with shingles cut from this tree, the turpentine which exudes from the wood, tending to make these roofs impenetrable to rain.

No one can fail to be struck with the curious catkins on the beech. The female blossom, which will become the beech-nut, is seated on the spray, whilst the male catkins hang down in clusters, shedding out their pollen upon every passing breeze.

The beech tree usually flowers every alternate year, so that possibly we may light upon a tree with leaf-buds only, and must then search further for another specimen bearing its catkins.

LARCH BLOSSOM.
Male.
Female.

The limits of space will not admit of a special notice here of other trees; but knowledge of the fact that this is the flowering season will lead to

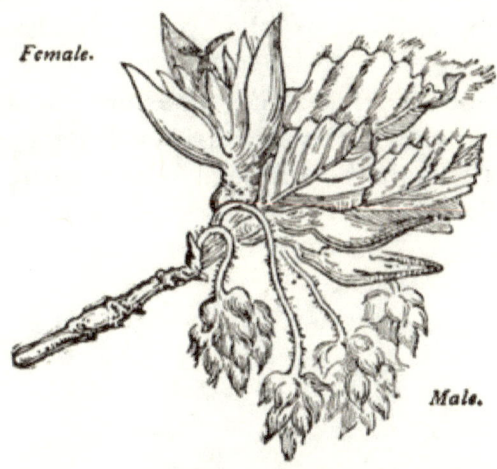

BEECH CATKINS.

some enjoyable study in hedgerows and woodlands. Let us not grudge some time and trouble spent in becoming acquainted with the inconspicuous but always interesting blossoms of our common trees.

IVY-LEAVED TOAD-FLAX (*Linaria cymbalaria*)

At this season the charming little ivy-leaved toad-flax may be found in the crevices of old walls, where its thread-like roots feed upon the decaying mortar. Penetrating deeply into the fissures of the brickwork, they both keep the plant firmly in its place and render it independent of cold, heat, and dryness. The cheery little plant holds its own and looks green and flourishing when prolonged drought is making other vegetation appear faded.

The winter and spring rains soak into the mortar of an old wall, and the horizontal roots of the toad-flax, protected as they are between the layers of bricks, have their store of moisture to draw upon and keep the plant in health and vigour.

If a root of this small creeper can be found within easy reach, it will repay a little careful observation through the summer. It possesses several points of interest besides the delicate beauty of its

IVY-LEAVED TOAD-FLAX.

tiny lavender and yellow flowers. It is closely related to the large snapdragons, but differs from them in having a spurred flower.

From its wonderfully prolific growth, this plant is popularly known as mother of thousands, and its drooping slender stems throwing a sort of veil over crumbling masonry must have given rise to its other familiar name of maiden hair.

The leaves are like miniature ivy, and when young are of a purple colour on the under side.

The chief interest in watching this plant is to observe its remarkable mode of sowing its seed.

As soon as the small capsule is formed it begins to turn towards the wall until it finds a crevice, and in that its places itself, just as we should put a small parcel on a shelf, and it remains secreted there until ripened by the warmth of sunlight, when the capsules split open, the seeds are shed out and lie upon the crumbling mortar, ready to germinate as soon as rain shall fall and afford them the needed moisture.

I often show this plant to my young friends as affording a remarkable instance of vegetable instinct and adaptation.

I am tempted to quote from Miss Ann Pratt's *Flowering Plants of Great Britain* an interesting incident connected with this humble flower.

In 1850 a deputation waited upon the Chancellor of the Exchequer respecting the abolition of the window tax.

A spray of *Linaria*, which had grown in the dark and produced only dwarfed and blanched leaves, was shown in contrast with another spray gathered from the same plant, which on its sun-lighted side was of a rich green and covered with flowers; this mute appeal was well calculated to show the evil and depressing effect of darkened dwellings and the consequent cruelty of the window tax.

Frog-Hoppers (*Aphrophora spumaria*)

> 'Insects of mysterious birth
> Sudden struck my wondering sight,
> Doubtless brought by moisture forth,
> Hid in knots of spittle white;
> Backs of leaves the burden bear,
> Where the sunshine cannot stray;
> "Wood seers" called, that wet declare,
> So the knowing shepherds say.'
>
> *Clare.*

The enjoyment of our rambles in woodland and garden paths is somewhat marred just now by quantities of a white frothy substance hanging on the grass stems, which clings to our clothes and is decidedly unpleasant. It has long been known by the name of cuckoo-spit, although it has nothing whatever to do with the cuckoo or any other bird. The French evidently credit the frogs with this production, as their name for it is *crachat de grenouille*, or frog-spit; but this is also wide of the mark.

If we examine some of the froth, we find a greenish white insect in the centre of each mass of it; this is a soft feeble creature, with small black eyes, the larva of the common frog-hopper.

The perfect insect will be found in summer by thousands in hayfields and pastures. It has marvellous leaping powers, as great in proportion to its size as though a man could spring four hundred yards up into the air.

The generic name *Aphrophora* means foam-bearing, in allusion to the little masses of froth in which the larvæ abide.

The female insect lays her eggs on grass stems, and when they are hatched the larvæ drive their probosces into the stems to obtain the sap on which they feed. The larvæ eject a quantity of white frothy matter, in which they live and by which they are protected, until they become chrysalides, and eventually emerge as brown frog-hoppers.

These creatures are closely related to the Aphides or green flies, and belong to the same order; only an aphis has four transparent wings, and these lively hoppers have the upper pair of wings opaque and hard, more resembling a beetle than a fly.

There are other branches of this family, such as the cuckoo-fly, which preys upon the succulent shoots of the hop-bine, the green frog-fly, which injures the potato, and many other species. They may all well be called pests, since they tend to mar the growing crops by sucking the juices of the plants, and from their activity and minute size are extremely difficult to exterminate.

June

'O June, O June, that we desire so,
 Wilt thou not make us happy on this day?
 Across the river thy soft breezes blow
 Sweet with the scent of bean-fields far away,
 Above our heads rustle the aspens grey,
 Calm is the sky with harmless clouds beset,
 No thought of storm the morning vexes yet.'
                                    William Morris.

June

THE LEAF-CUTTER BEE

BESIDES the common honey-bee, we possess in England many hundred species of what are called solitary bees. Their lives are extremely interesting, for many reasons. They live in all sorts of places, some in holes in our gravel walks, some in dry banks, where they form long, deep burrows in which they lay their eggs, and then close up the holes, leaving the young bees to find their own way out. Other species adopt ready-made holes in walls and brickwork, in which to rear their families. Empty snail-shells may often be found half full of dried mud placed there by one of these eccentric bees, and if we examine this deposit we shall find small cells, which are the cradles of the immature bees. A hollow bramble stem is the choice of the Mason bee (*Osmia leucomelana*). In this convenient circular chamber the bee sets to work and removes some of the pith till she has a clear space of five or six inches; then, having prepared and masticated some substance which she knows to be suitable for the food of her grubs, she places a small quantity of it at the end of the hollow space and lays an egg in it, so that when hatched the larva will only have to feed

LEAF-CUTTER BEE AND ROSE LEAF
(Natural size).

and grow till it changes to a chrysalis. In that condition it remains through the winter, and comes out a perfect bee in the following June. Six or eight eggs are thus laid in one bramble stem, each divided by a thin partition.

I constantly see another of these very curious solitary bees at work on my rose trees. She is known as the Upholsterer bee (*Megachile centuncularis*), so called from her dainty fashion of lining her nest with rose leaves. The nests are not easily found, but I was fortunate enough to light upon a specimen, and could examine its curious formation.

The bee settles on the edge of a rose leaf, and holding it firmly between its forelegs, saws out a round piece of the leaf and flies away with it. About ten or eleven of these pieces are required to line the burrow the bee has scooped in the bank; they are neatly fitted together without any sort of cement, and as they dry they curl up and form a neat little tunnel. In this the bee stores up the honey and pollen of thistles, which form a sweet and suitable food for her infant bees. When a sufficient number of eggs has been laid in the tunnel, the end is securely closed up with three pieces of leaf neatly joined together, and then, her work being completed, the mother flies away and leaves her nursery to manage for itself.

Some of the solitary bees are smaller than house flies, others are as large as humble bees; some are jet-black, others are yellow or brown. They flourish in great variety through the spring and summer months, and their remarkably interesting habits should lead young people to inquire about them.

As a guide in identifying the various species, I

would recommend *British Bees*, by W. E. Shuckard (published by Lovell Reeve & Co.). With this book, a small net, and a magnifying glass, I can promise my readers a very fascinating pursuit for their summer rambles.

The bees may be found on flowers, gravel walks, turf, old walls, and hedge banks; they are easily caught, and can be kept under a glass until we have ascertained all we desire to know about them. Then we may set them at liberty, as we shall have learnt the appearance of each species, and can recognise them as we see them busily at work out of doors.

Unless a dried collection of insects is really needed for scientific purposes, I always strongly discourage the indiscriminate killing of insects; it seems to me that it must tend to blunt kind and tender feelings in young people, and it is really needless, except for those who are in training to become practical scientists.

THE HOVERER-FLY (*Syrphus plumosus*)

As the humble-bee fly is a harbinger of spring and one of the first insects we may see visiting the early blossoms of the year, so the hoverer-fly betokens the arrival of summer. It revels in the hottest sunshine, and is one of the most active, swift-winged creatures imaginable.

The specimen I watched to-day was a *Syrphus plumosus*, one of the handsomest of the species. It is covered with yellow down, the wings having a few dark markings, and its general appearance is so like a small humble bee that most people would take it to be one.

This fly seems quite as intent upon studying me as I am to learn about it; it poises in the air for a minute or two, staring at me, humming loudly and watching my every movement. It is quite curious to observe how stationary in the air the creature remains, its wings quivering with such exceeding rapidity that they are quite invisible, so that one is puzzled to imagine how the insect is supported in the air. Thus it will remain until I make some slight movement, when instantly the fly is gone, and my eye cannot trace its flight.

One day I desired to make a drawing of a *Syrphus*, and I shall not soon forget what an exercise of patience it

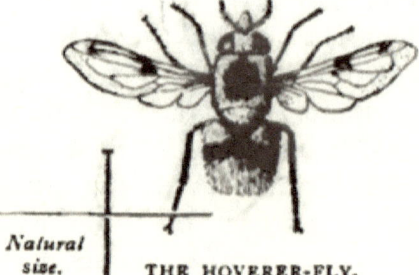

Natural size.   THE HOVERER-FLY.

was to capture it. I did succeed at last by a quick sweep of a gauze net, and my captive was detained for a while until I had taken its portrait. It had not the patient gentleness of the humble-bee fly, but continued to buzz and fuss in an angry manner until I was able to set it at liberty.

ICHNEUMON FLIES

If we observe creeping up the window panes or hovering over the flower beds some curious-looking flies, with very slender bodies and antennæ constantly quivering, we may know them at once to be ichneumon flies. They have a strange and cruel habit of laying their eggs in living caterpillars and chrysalides, and they are ever on the watch to find some unfortunate insect which shall become a receptacle for their progeny. These flies are of all sizes, ranging from a minute creature like a small gnat up to the one figured in the illustration. When we consider that almost every

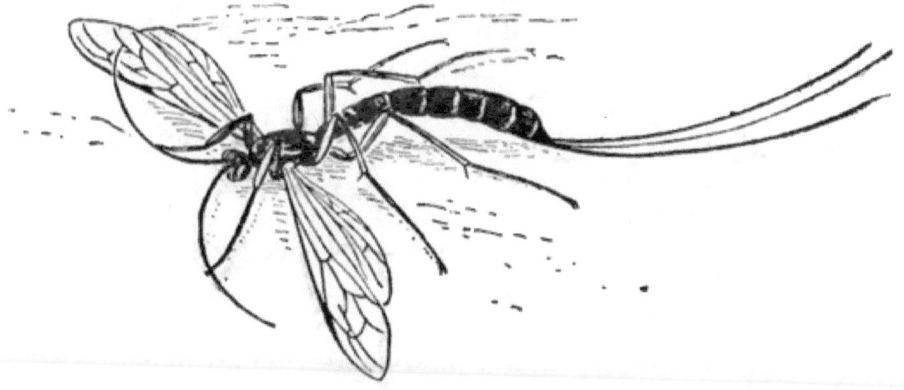

RHYSSA PERSUASORIA.
(*Parasitic upon wood-boring larvæ.*)

insect has one (or more) enemy of this kind, we may imagine that ichneumons abound in our gardens, and when once our attention has been called to them, we shall quickly know them by sight. They are peculiarly restless insects, always prying into flowers, and running up and down the leaves in a never-ending search for their prey. They are doubtless of great use in keeping down the hosts of caterpillars that feed upon our vegetables; and all through the spring and summer months this secret warfare is going on.

The ichneumon fly is furnished with a long thin ovipositor, which enables it to pierce the skin of the caterpillar and deposit a number of eggs in its body; these hatch into minute grubs, which feed upon the internal organs of the caterpillar. The victim does not appear to suffer; it goes on consuming its food and growing until the ichneumon grubs are nearly mature. They then attack its vital organs until the caterpillar dies, and the grubs, after turning into chrysalides, hatch into the perfect insect.

I well remember my surprise and disappointment some years ago when a caterpillar, from which I expected to rear a very beautiful moth, instead of turning into a chrysalis, suddenly became covered with small yellow cocoons, which, I need not say, presently turned into an unwelcome swarm of ichneumon flies.

It was in this way that I first made the acquaintance of this tribe of insects; and ever since I have been learning the immense variety of species which exist, and their subtlety in pursuing their prey.

FLAX (*Linum usitatissimum*)

The delicate pale blue flowers of the flax are now opening, and remind me afresh, not only of the beauty of this plant, but of its great usefulness also. We owe to the strong fibres of its stem our linen, cambric, lawn, lace, and thread; its seeds, when crushed, produce the valuable linseed oil so much in use by artists, and required in house painting and in many trades and manufactures.

We all know the remedial effects of linseed-tea, and the value of the ground meal which forms soothing poultices to relieve inflammatory pain; and finally when the oil has been pressed out of the seeds the mass of husk which remains is made into cakes, which form an excellent and fattening food for cattle. Surely we ought to look upon such a plant as this with admiring gratitude as we remember its many uses.

Those who possess a garden or even a few pots upon a window ledge can easily grow their own flax plants by sowing a pinch of the seed in May. It only needs good soil and watering, in order to produce an abundance of its delicate blue flowers, and when they are over we can see for ourselves the round seed capsules, like little balls, which are alluded to in Exod. ix. 31: 'The barley was in the ear, and the flax was bolled,' that is, swollen.[1]

This leads us to reflect upon the great antiquity of this plant, and its frequent mention in Scripture.

It is believed that flax has been cultivated in Egypt for five thousand years, and great quantities of it must therefore have been grown, to supply the immense demand for mummy cloth, as it was invariably made of linen, either fine or coarse.

From a reference in Ezek. xxvii. 7, we learn that sails for ships were made of linen; which again shows that the fibre could be woven either into the finest cambric or a cloth of the coarsest and most durable nature.

When the stems are mature they are dried and split, then steeped in water, and afterwards hackled into

---

[1] This derivation is taken from Professor Skeat's *Etymological English Dictionary*.

threads by means of a comb, to separate the coarser fibres and leave the fine strands which are fit for weaving purposes.

The severe processes required to make the stems into material for the loom have led to the flax-plant being used as an emblem of the Divine puposes of earthly trials. The Venerable Bede thus speaks of it: 'The flax springs from the earth green and flourishing; but through much rough usage, and with the loss of all its native sap and verdure, is at last transfigured into raiment white as snow; thus the more that true holiness is tried and afflicted the more brightly does its beauty come forth.'

I must add one other thought in connection with this plant. The simple little lamp of sun-dried clay used by the village people in Palestine, is filled with olive oil, and burns by means of a few fibres of flax

FLAX.

inserted as a wick. When the supply of oil becomes exhausted this flax-wick gives out a pungent smoke, so that either more oil must be added or the lamp extinguished.

In Isa. xlii. 3 we find the promise, 'The smoking flax shall He not quench,' referring to the infinite mercy of the Saviour, who will cherish the least spark of grace in the human heart and foster it until the dimness passes into a shining light.

The flax of commerce is extensively grown in Ireland, to supply material for the manufacture of linen; it is also cultivated in some parts of Scotland, and may be found growing wild in fields and waste places in England, but it is not truly indigenous.

Our only British species of flax is *Linum catharticum*, or white flax. A graceful little plant growing about five inches high with small drooping white flowers. It is said to be violently poisonous.

THE SNAKE-FLY (*Raphidia Ophiopsis*)

The hot weather we have lately had has driven quantities of different kinds of flies indoors. Certain passage windows on the north side of this house are full of interest for me, as I find there quite an assortment of winged creatures, not the common house fly, but large and small ichneumons with their ever-quivering antennæ, minute gall-flies, brilliant green and golden sun-flies and a host of others whose names and life-histories are as yet unknown to me.

One very remarkable four-winged fly appeared amongst the throng about ten days ago It struck me as being rather rare, so I placed it in a glass globe, in order that I might become more intimately acquainted with its habits and manners.

After some little searching amongst my books I found that I had captured a snake-fly, a most appropriate name for a creature with a long slender neck and a flattened, vicious-looking head, which at once suggests the idea of a viper.

This fly is a highly sensitive little creature, it is

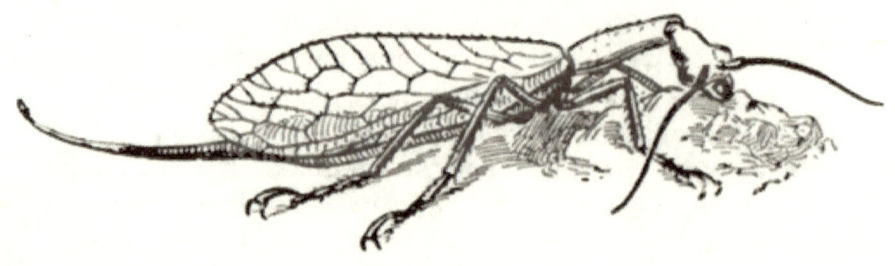

SNAKE-FLY AND LARVA

(*Snake-fly, five times nat. size. Larva, six times nat. size*).

on the alert the moment it sees or hears anything unusual, lifting up its little serpentine head, and glancing from side to side with much intelligence; full of courage, also, for it will try to seize a small twig or anything held near it. Although its natural food consists of small insects, I find the snake-fly will eagerly accept little morsels of raw meat, upon which it fastens its powerful mandibles, and with a lens I can watch it evidently enjoying a hearty meal.

Almost all flies are fond of sugar and honey,

so I offer these dainties as well as meat, and they appear to be highly approved. The larva of this fly lives under the bark of trees, where it, as well as the fly, feeds on minute insects.

My specimen possesses a long ovipositor, which gives it a rather formidable appearance; but I do not imagine that this organ could ever be used as a weapon of offence or defence; it is simply a long tube, by means of which the insect is able to deposit its eggs in suitable crevices in the bark of trees. I had written thus far the description of my snake-fly when, a few days later, whilst sitting under a tree, there dropped upon the book I was reading a wriggling creature, which I saw at a glance must be the larva of the said fly. There was the snake-like head, the long neck and slender body, only no wings and no ovipositor.

I secured my captive, and supplied it with some raw meat, which it pounced upon and devoured with avidity. Whether I can succeed in keeping the creature until it turns into a perfect snake-fly remains to be seen; at any rate, I shall keep and feed it, hoping it may prosper in my hands.

Growing Seeds

Those who do not possess a garden may like to know how much interest and pleasure can be derived from a window box, or even a few pots full of earth. If we like to have only colour and perfume, it is easy to select the plants we prefer, and either grow them from seed or by means of slips; but I would suggest some other modes of turning the small space to account, some very simple experiments which will be both interesting and instructive.

The early growth of a seed is seldom observed, because the process goes on underground, and therefore out of our sight; but by planting about a dozen broad beans in moist earth and taking them carefully up from time to time, we shall then see how they become plants. In a week's time we shall find that out of the dark spot on the bean a root is pushing out into the soil; the tip is protected by what is called a root-cap; it has to make its way through stones and earth particles, and as it wears away it is always being renewed from within. A little later we shall find that a shoot, called a *plumule*, is making its way up to the surface to reach the light. The growing point of this shoot is very tender, and as the existence of the plant depends upon its being uninjured it is protected in a curious way. The *plumule* is curved, as shown in the drawing, and the bowed part of the stem does the rough work of pushing through the earth until it reaches the surface; then the stem straightens, and the *plumule* grows on into a young bean-plant.

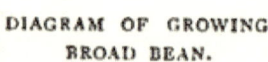
DIAGRAM OF GROWING BROAD BEAN.

A horse-chestnut and an acorn will show the early growth of two of our great forest trees. These can be grown very well in damp moss, which will enable us to watch the growth of the first shoot, until it divides into 'the branch which groweth upward and the root which groweth downward.' Tamarind seeds taken out of the preserve will grow readily, and have an amusing way of throwing off their seed-coats, and appearing on the surface of the soil like ivory buttons, and soon begin to show finely divided leaves of a most delicate green tint. Orange and lemon pips develop into charming little ever-green trees, and are well worth growing, as they will live for years in small china pots and form useful table ornaments. They require frequent sponging, to prevent dust from clogging the leaves, and the stems must be cleared of the brown scale insect, which sometimes attacks and injures orange and lemon trees.

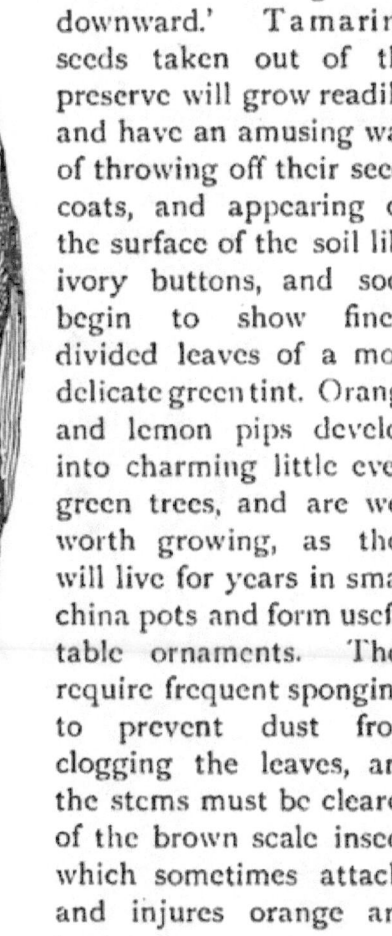

PODOPHYLLUM BUD.

SANGUINARIA.

When our attention is called to any particular branch of nature's handiwork we are sure to go on making discoveries. Watching the growth of seeds led me to study various plants in their early stages, and I have had two extremely curious instances of leaf-folding recorded for my readers' benefit, as they are unusual plants not likely to be met with in ordinary gardens.

In early spring the podophyllum sends up sturdy dark-brown shoots with a knob at the top, which afterwards develops with the flower, and the leaf is wrapped like a mantle around the stem. It has a ludicrous resemblance to a shivering form enveloped in an Inverness cape!

The other plant, sanguinaria, possesses a delicate white flower, which is tenderly protected by an enfolding leaf sheltering it until it is fully expanded. These suggestions will, I hope, direct attention to the endless beauties of unfolding plants in early spring.

July

'That sunset! look beneath the boughs,
Over the copse—beyond the hills;
How soft, yet deep and warm it glows,
And heaven with rich suffusion fills;
With hues where still the opal's tint,
Its gleam of prisoned fire is blent,
    Where flame through azure thrills!'
          Currer Bell, or Charlotte Brontë.

July

Amber

I HAVE in my museum a piece of amber in which some small flies with gauzy wings can be plainly discerned. Ages ago these insects must have alighted upon some resin oozing out of a pine tree, of a species that is now extinct (*Pinus succinifer*), and, held fast by the glutinous sap, they were embedded and enshrined there, until, in the course of time, the resin became mineralised into what we call amber.

Although this substance is occasionally found in England and France and rather plentifully in Australia, the chief supply comes to us from the south-eastern shores of the Baltic. A forest of the amber-yielding pine must have existed there long ago. It is now submerged, and in calm weather the fossil trees and immense deposits of amber can be discerned on the ocean-floor.

The amber fishers, clothed in leather and provided with hooked forks and hand nets, wade into the sea, and gather such fragments of amber as may be floating on the surface; but the larger and finer pieces are obtained by rowing out from the shore and raising the masses of amber with pronged forks and nets. Even better results are obtained by divers, who work under water for five hours at a time, prising up large blocks of amber from the weed and sand in which they are embedded; these are hauled up to the boats and brought to shore.

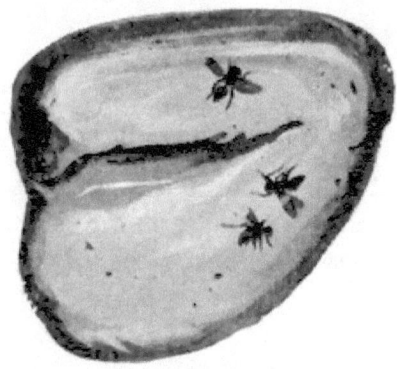

FLIES IN AMBER.

Amber is chiefly used for mouth-pieces for pipes, partly because of its smooth surface, and originally on account of the belief which prevails in Turkey that it cannot transmit infection. Some amber, like my own specimen, is as clear as yellow glass, while other pieces are more or less clouded. The first mention of this substance is in Homer's *Odyssey*—

'An artist to my father's palace came,
With gold and amber chains.'

So we learn that necklaces of amber are of high antiquity.

As many as eight hundred different kinds of insects have been discovered embedded in amber, all formerly natives of warm climates, but now extinct.

If a piece of amber is firmly rubbed upon flannel or cloth, it will become so electric as to attract small pieces of paper, which will adhere firmly to it. To this electric quality we may trace its Greek name of *electron*, from which our word electricity is derived.

If we like to experiment with a piece of amber and apply it to a candle, it will burn, giving out a rather disagreeable odour and black smoke; but if we blow out the flame, there then arises a white vapour, which exhales a pleasant aromatic scent.

To this Milton refers when he says in *Samson Agonistes*—

> 'An amber scent ot odorous perfume
> Her harbinger.'

As may be gathered from numerous references in our old poets, the aroma of amber was used in the Elizabethan age to give gusto to foods and wines as well as to perfume garments.

Rocks and Stones

It has always been a source of interest to me to observe the various kinds of stones I meet with in a morning's ramble. Living, as I do, where quartzite pebbles abound, I am always being reminded that the sea once covered this place, although it now stands between 400 and 500 feet above it, and that it was by the sea's action that these stones were rolled backwards and forwards, until all their angles were smoothed away. In fact, they are exactly such as we may find on any sea beach at the present day.

> 'Where rolls the deep, there grew the tree;
> O earth, what changes hast thou seen?
> There, where the long street roars, hath been
> The silence of the central sea.'

Common flints out of a chalk-pit are usually dark grey or black within the outer white crust; but our quartzite flints are beautifully stained, banded, and veined, and partake of the nature of agate and cornelian. When polished they form ornamental paper-weights. Red jasper, fit to be cut into seals, is also abundant here.

Blocks of pudding-stone are occasionally exposed in our fields as the plough turns up the soil. This stone was once grey mud, into which pebbles large and small became embedded; then, in process of time, the mass hardened into solid rock, which, when

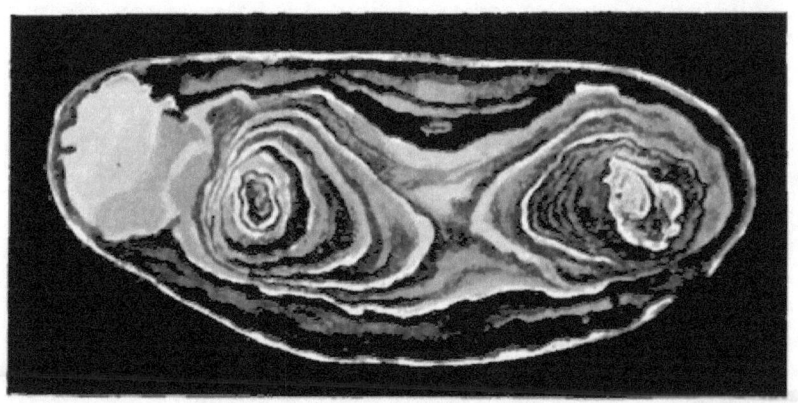

POLISHED QUARTZITE PEBBLE.

sawn into pieces, will take a fine polish, the stones in it looking much like plums in a pudding, hence its common name.

Some of my readers may live in mountainous places where granite rocks exist; these will afford an interesting subject for study. Granite consists mainly of three substances, the white or yellowish grains being quartz, the pink felspar, and the black mica.

In the Museum of Geology[1] in London we can see

[1] This museum in Jermyn Street is always open and quite free of access.

case after case filled with specimens of polished granite of every description, and of great variety of appearance. That which is mainly felspar is bright pink or dark red; some pieces are light grey as quartz predominates, and the darker kinds are full of mica.

I always glance at the heaps of stones by the roadside, since a very slight knowledge of geology tells me where they are likely to have come from, and an otherwise uninteresting walk along a dusty

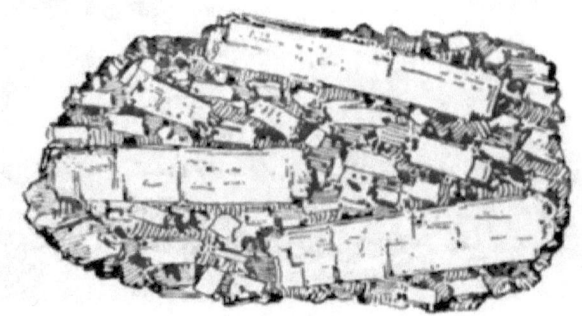

CORNISH GRANITE
(*showing orthoclase crystals*).

road may be enlivened by a little thought about the materials of which the road itself is made.

Even the kerbstones of the London streets, when washed clean by a heavy shower, reveal by their varied tints of grey, red or pink, that they have come from quarries in Scotland, Cornwall, Devonshire, or the Isle of Man.

In the beautiful Cornish valley of Lamorna, blocks of granite measuring twenty-five feet in length by eleven feet in diameter have sometimes been cut, and the plinths for the railings of the British Museum came from the Carnsew quarries in Cornwall.

Should any of my readers pay a visit to the Land's End, they will be able to observe in the curious columnar granite blocks on that coast the pieces of felspar (of the variety called *orthoclase*), sometimes as much as three inches in length, which give this granite a very distinct character.

It is quite worth while to know something of the nature of the country in which we may happen to live; to learn, for instance, whether the soil is gravel, chalk, clay, or sand. I am often surprised to find young people unable to answer an elementary question upon this point, because they have never given any thought to the subject.

In some places it is easy to see at a glance of what the soil consists, every hedge-bank displaying either clay, stones, or chalk, as the case may be. Other places, especially on level ground, grass-fields, and arable land, do not reveal much about the nature of the subsoil.

Railway cuttings, gravel pits, and excavations are aids to a knowledge of the soil which lies beneath the surface, and clay has an unpleasant way of insisting upon making itself observed, in the miry footpaths which make our walks so tiring in the winter months.

These remarks may set some students thinking upon the simple problems of geology, to which I hope to return in next month's ramble.

WILD TEASEL (*Dipsacus sylvestris*)

If my readers can find a specimen of wild teasel growing in some hedge-bank they will, I think, be interested to hear a little about its structure and uses. It is a striking-looking plant, growing from four to six feet high with a straight stem and opposite leaves, which have the peculiarity of uniting at the base so as to form a cup-shaped receptacle, holding nearly half a pint of clear water. Into this liquid small insects fall and become decayed; the wind also blows dust and dead leaves into the water, so that in time it becomes rich in organic matter. This is absorbed by the plant, and tends to nourish and strengthen its growth.

These leaf-basins also serve another purpose. It is necessary that the flowers should be fertilised only by winged insects, and there seems little doubt that the water retained at the base of the leaves tends to isolate the central stem, and thus snails, slugs, and ants are prevented from crawling up to the flowers.

TEASEL.

The common cow parsnip has huge inflated sheaths at the base of its leaves, which contain water, both to nourish and protect the flowers in a similar manner.

I have not met with the smaller species called fuller's teasel (*Dipsacus fullonum*). It is cultivated in some parts of England, and very extensively abroad in France, Austria, and other parts of Europe.

I read in the *Treasury of Botany* that in 1859 we imported from France nearly nineteen million teasel-heads, valued at five shillings a thousand. The bristly seed-vessels are employed by manufacturers to raise the nap of cloth. The capsule consists of very sharp elastic points hooked at the end, and when rows of these spiky balls are affixed to a small wooden frame they form a kind of currycomb, which when drawn over the surface of woollen cloth raises up a soft nap.

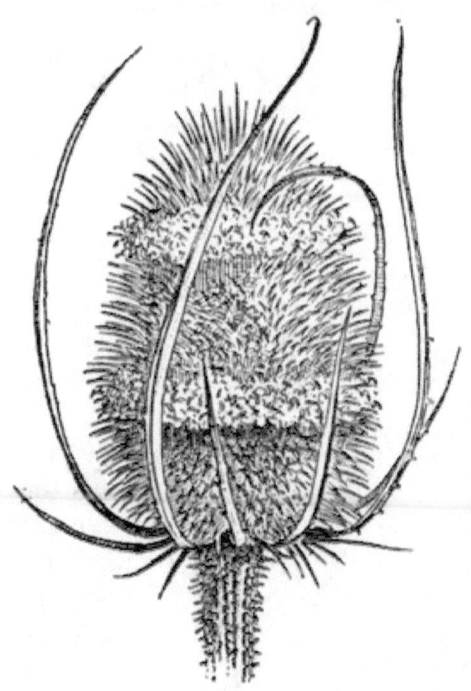

TEASEL HEAD
(*Natural size, to show bands of florets*).

The process has been imitated by machinery, but the fuller's teasel is, I believe, still extensively used.

The poet Dyer alludes to this useful plant.

> 'Soon the clothier's shears
> And burler's thistle skim the surface sheen.'

A burler is a man who pulls out the 'burls,' or small knotted lumps in wool or thread.

One other characteristic of the teasel is worthy of remark. The bristly flower-head expands its florets regularly. First a band of pale lilac will appear about the middle; when that withers a row of florets above and then one below will expand; but never can we find the handsome flower-head all expanded at once. It cautiously opens a little at a time until the insects have done their work and all the florets have been fertilised.

WILD SUCCORY.

WILD SUCCORY (*Cichorium tutybus*)

Some plants seem to have a strong preference for dry, dusty roadsides and footpaths. The plantain, for instance, never flourishes more vigorously than on a well-trodden path, and the wild succory is another plant so associated with roadsides that the Germans call it 'keeper of the ways.'

When growing wild, succory presents little beauty in its leafage; its stiff, wiry stems spring up out of the hard chalky soil which it prefers and into which it sends down a long tap-root in order to collect all possible nutriment and moisture. Even its lovely sky-blue flowers have a tantalising way of growing without stalks, one here and one there, scattered along the stem, so that we cannot form a bouquet of them; and almost as soon as they are gathered they close up, before we have time to admire their beauty. They need not, however, be thrown away, for they will expand again in water if placed in sunlight.

Succory takes its place among the flowers included by Linnæus in his floral clock, formed of such plants as opened and closed their blossoms at certain hours of the day. It is an early riser, and greets the morning sun with its star-like flowers between four and five o'clock.

> 'On upland slopes the shepherds mark
> The hour when, as the dial true,
> Cichorium to the towering lark
> Lifts her soft eyes serenely blue.'

As if to make up for this early blossoming, the petals begin to close between nine and ten in the morning, and the plant sleeps for the rest of the day.

I have given the hours as observed by Linnæus at Upsala. They are probably different in England, and on cloudy days the flowers scarcely open at all.

July

Some years ago I dug up a root of wild succory and had it planted in my garden in good soil and in a sunny aspect. In the course of years it has amply repaid me by growing into a sturdy plant three or four feet high; and in this month, when it is always covered with its star-like, exquisite blue flowers, it forms one of my cherished garden treasures.

Chicory or succory is largely grown for the sake of its tap-root, which, when dried and ground, is used to mix with coffee.

The endive we use for salads is an allied species, a biennial plant derived originally, I believe, from *Cichorium pumilum*, a wild plant still found commonly along the shores of the Mediterranean.

Fishing for Insects

Towards the end of this month every little stream abounds with insect life. Of this there may be no appearance on the surface, but a few sweeps with a muslin net will bring to light a variety of interesting creatures.

Provided with a canful of water and a net, I went off this morning to my pond to see what I could discover. Passing the net through some water-weed, I was not long in finding greyish-green beetles leaping vigorously in my net; these were the water-boatmen (*Notonecta glauca*). The body of this insect is shaped just like a boat, and the two long hind legs with which it propels itself are feathered like oars. This beetle swims on its back, and spends much of its time resting on the surface of the water, diving now and then to catch some insect on which it feeds.

I hardly liked to touch a sluggish crawling grub which was burying itself in the mud that I had brought up in my first haul from the pond. This creature, however, proved well worth examination, for it was a dragon-fly larva, provided with a remarkable lobster-like claw with which to seize its prey. As the grub lies concealed in the mud some insect approaches it, and as soon as its prey is within reach, the claw, which has been folded up out of sight, is darted out and secures the insect with unerring aim.

DRAGON-FLY PUPA
(*Natural size*).

I was presently fortunate enough to secure several of the larger kind of beetles, and amongst them *Dytiscus marginalis*, the male possessing smooth brown wing-cases and the female having furrowed elytra. The curious discs upon the fore-legs of these insects are worthy of notice, for they possess the function of suckers, and enable the beetle to fix itself firmly to any solid substance.

Securing one of these beetles in a jug, I tried to pour it and the water into a globe, but the beetle, to my surprise, remained at the bottom of the jug, holding itself firmly there by its suckers.

DYTISCUS MARGINALIS AND LARVA
(Natural size).

The larva of this genus is well named water-lion, for it is fiercely voracious, preying upon all kinds of other water insects, and has a peculiarly repulsive aspect, with its pair of curved cruel-looking jaws and flat snake-like head. When I tease it with a piece of twig it flies at it, and will defend itself when attacked with a dogged sort of courage. It always seems to me like a shark amongst the milder inhabitants of the pond.

I must not be tempted further to describe what our net brings to light. If we place our captures in a globe of water, and then read about them in some handbook to natural history, we shall not fail to learn many interesting facts about the curious habits of the creatures inhabiting our ponds and ditches.

August

'The quiet August noon has come;
A slumberous silence fills the sky,
The fields are still, the woods are dumb,
In glassy sleep the waters lie.
And mark yon soft white clouds that rest
Above our vale, a moveless throng;
The cattle on the mountain's breast,
Enjoy the grateful shadow long.'

Bryant.

August

The Great Green Grasshopper
(*Acrida viridissima*)

AS the fields are now teeming with grass-hoppers, large and small, it will be quite easy and well worth while to capture a few, and note their curious form and varied markings. Those we find in the meadows are usually of the same tint of green as the grass on which they feed; but if we collect these insects from a bare chalky soil, they will be grey-coloured, so as to imitate the general tone of the ground they rest upon.

There is also a very handsome species, which is a tree-dweller, and may be found at this season in some localities by shaking oak branches; in other places I hear of their being caught in hazel hedges and on sunny banks, where they are easily secured with a small butterfly-net.

I kept a specimen of this insect a few years ago, and find it a very interesting pet. A glass globe covered with a piece of net forms a suitable home for it, and although it prefers flies and small insects, it will eat raw meat and succulent cabbage-stalks.

No one could fail to admire the exquisitely brilliant green of this insect with its golden eyes, and its long delicate wings, which, however, it does not seem to use except when they are expanded to break the force of its fall from tree branches. The antennæ are long and tapering, and my specimen, being a female, possesses an extended ovipositor.

This species measures from two and a half to three inches from head to tail, and taking into

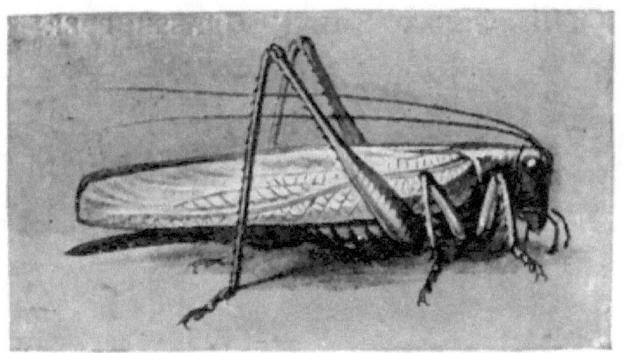

THE GREAT GREEN GRASSHOPPER.

account its size and brilliant colouring, it is perhaps one of the most striking of our British insects.

Its habits are very dainty, every speck of dust is at once removed from its legs and feet; the tapering antennæ are drawn through its feelers, and they also cleanse the delicate wing-cases. In fact, as one watches all this going on, one is led to wish that all human beings could be persuaded to learn from this lowly creature to perform their toilets as carefully.

BIRD'S-FOOT TREFOIL (*Lotus corniculatus*)

Seeing the sparrows busily feeding upon the seed-pods of the bird's-foot trefoil, which grows much too freely upon my lawn, I have been led to-day to reflect upon the great value to wild birds of the various weeds which cover every piece of waste ground.

The many weeks of dry weather we have had this summer have brought the birds almost to starvation point. The lawns are hard and cracked with the continual sunshine, so that the thrushes

BIRD'S-FOOT TREFOIL.

and blackbirds can find no worms or slugs, and very naturally they resort to the fruit gardens in the absence of other food.

The mountain ash and elderberries are also eagerly sought for and devoured, and then weeds are resorted to, and keep the famished birds alive until the welcome rains restore their accustomed insect diet.

Few people seem to know that wild birds need feeding quite as much in a long dry summer as in a hard winter, and a pan of water is also a great luxury to our feathered friends. All kinds of finches feed greedily upon thistle seeds, and many other species seek for their favourite chickweed and groundsel, plantain, vetches, and hawkweed.

Other weeds are the resort of shy birds that we seldom see in the act of feeding, because their keen sight and hearing give them warning of our approach, and they slip away under cover until we have passed by.

Wild pigeons, if they do a great deal of harm in eating more than their share in the corn-fields, also do some good by feeding upon charlock or wild mustard, one of the most troublesome weeds the farmer has to contend with. They also eat the seeds of various polygonums which are sure to abound in fallow land.

We see then that weeds are really wayside provisions for the feathered tribes, and fulfil an important office in maintaining their lives when other resources fail.

The illustration shows the resemblance between the trefoil pods and a bird's foot, hence the appropriateness of its name. It is a happy time for the humble bees when this plant, with its pretty yellow blossoms, is out in flower—the lawn is so covered with the busy little insects one can hardly walk without treading upon them.

HOME-MADE INK

A curious fungus known as the maned agaric (*Coprinus comatus*) is now growing in abundance in a grassy nook behind some evergreens, where it

August

always makes its appearance in the course of the autumn.

It is like a cone-shaped mushroom of snowy whiteness, with a few brown specks on the upper part of the cap. When the stem is four or five inches high, the lower part of the cap becomes fringed, and begins to drop a jet-black liquid, which creates a dark ring upon the ground. So long ago as August 1888 it struck me that this liquid might be utilised, and accordingly I tried an experiment in the manufacture of ink. The agarics were placed in a basin over night, and by the next morning I found they had melted into a quantity of ink as jet black as I could desire. The lines I wrote with this liquid are as bright and clear to-day as they were when first penned eleven years ago.

COPRINUS COMATUS.

The only preparation needed is that the ink should be boiled, strained, and then have the addition of a little corrosive sublimate, to prevent any fungoid growth. The specimen bottleful I made in 1888 has remained clear and usable to this day.

It is singular that a substance so exquisitely white as this fungus is in its early stage should when melting away become of such an inky blackness. It is a circumstance about which I can offer no explanation.

Blue Butterflies Asleep (*Lycæna Icarus*)

Walking last evening in a field where the long flowering stalks of grass were swaying to and fro in the breeze, I was struck by what seemed a small grey blossom hanging upon one of them, and looking more closely I found it was a blue butterfly which had gone to sleep upon the grass stem. Passing on a little further, I found dozens of the exquisite little creatures with folded wings quietly resting until the sunrise should awaken them to new life and activity.

This morning there was heavy rain and a high wind, and I was rather curious to know how the butterflies had fared; so when there came a lull in the storm I made my way to the field, and there were the fragile little insects being blown hither and thither on the grass stalks, but evidently quite unharmed by wind and rain.

I could but admire the instinct which had guided these frail creatures in their choice of a resting-place; had they been roosting in trees or shrubs, a blow from a large leaf flapping to and fro would have been fatal to them, but on the slender grasses they bent before

COMMON BLUE BUTTERFLIES REPOSING ON GRASS.

the gale and swung in their aerial cradles quite unharmed.

Another point of interest is that the bright azure of their upper wings, which would have made them a conspicuous mark for a passing bird to feast upon, was entirely concealed whilst they were thus at rest, the wings being closely folded and bent down, so that the finely spotted under-wings alone were seen, and made the tiny butterfly look like a part of the grass itself.

The calm confidence of these pretty insects brought to my mind a saying of Martin Luther, as he called attention to a young bird asleep upon a spray. 'This little fellow has chosen his shelter, and is quietly rocking himself to sleep, without a care for to-morrow's lodgings, calmly holding by his little twig and leaving God to think for him.'

Varying Position of Leaves

A long period of drought is now rather seriously affecting vegetation. Without moisture, the roots of plants cannot send up the needful supplies of food into the stem and leaves; exhaustion consequently ensues, and the outward sign of a starved condition is seen in the drooping position of the leaves.

Where the leaf-stalk joins the stem there is a flexibility of tissue which admits of the leaf being raised or lowered. In some trees and plants there exists at the base of the leaf stalk (or petiole) a swollen articulation which is called a *pulvinus*. It it almost like a hinge, and enables the leaf to hang down or rise to an entirely upright position.

We may see this hinge in action by touching a

sensitive plant, when before our eyes the leaf rapidly descends and the leaflets fold together.

Where this plant cannot be observed, the same effect can be noted by examining a clover plant in the morning, when all its leaves will be erect; and visiting the same plant in the evening, each leaflet will be found hanging down and folded together in its nightly sleep.

The illustration shows the effect of drought upon

RHODODENDRON.

rhododendron leaves. This pendent foliage has a strangely depressing effect upon the spirits; it is as though all nature was sorrowing, and trying to express her mournful condition.

As far back as the time of the Egyptian dynasties, the upward tending line was always chosen as the expression of joy and gladness, typified by a man with uplifted hands, that being the hieroglyph to express rejoicing. The upward curves of a smiling

EGYPTIAN HIEROGLYPH FOR REJOICING.

mouth, and the sad effect when the lips are drawn downwards, illustrate the same truth. For the same reason we call a tree whose branches all droop towards the ground a weeping willow, birch or elm, as the case may be.

Keeping this principle in mind as we take our rambles will afford a fresh subject for thought, and we shall find many other illustrations confirming this fact, which I have not space to touch upon now.

Rocks and Stones

In a previous note I spoke of some points of interest in the formation of granite rocks, and what we may discover in gravelly soils. Let us now suppose ourselves in a limestone country, with its granite cliffs and caverns.

It was a delightful surprise to me to find that I could actually pick up fossils in the streets at Buxton, which are mended with broken limestone; I thus obtained quite a variety of museum specimens in the course of a morning's walk. There are, I believe, more than six hundred species of fossil shells to be found in mountain limestone, besides the remains of fishes, corals, and plants.

Derbyshire abounds in curious caverns, where we may see the growth of stalactites from the roof. These are formed by the constant dripping of water containing calcareous matter, which encrusts into long spikes like icicles. The drops continually falling from them also concrete upon the floor of the cavern, and form masses of what is called stalagmite.

I met with a still more curious form of this deposit in a cavern at the Cheddar Cliffs. The dripping lime water had there taken the form of a curtain, and hung from the roof in graceful folds; it was so translucent that the light of a torch, held by the guide, shone through as though it were formed of horn or tortoise-shell.

Alabaster is another form of limestone; this is a sort of calcareous spar, soft enough to be easily carved into statuettes and other ornaments.

Some years ago, when I was visiting a little seaside resort called Blue Anchor in Somersetshire, I was much interested in observing that a part of the sea cliff there contained a vein of alabaster of various shades of pink and red. Although it is found in many places in England in strata in the earth, or in caverns, I do not know of any other locality where alabaster can be seen and obtained so easily as at this particular spot.

As I am only trying to point out a few interesting geological specimens which my readers may find for themselves, I will pass over the various kinds of marbles—forms of limestone which need to be quarried out of the earth, and which are seldom to be met with in a day's ramble.

Where building operations are going on we may often obtain small pieces of the Bath, Portland or Caen stone, which are used so much for pillars and ornamental sculpture.

The additional names of oolite and roe-stone have been given to these forms of limestone, because they appear to consist of small round grains or eggs, such as compose the roe of a fish.

Palestine Oaks

When we read of the oak trees mentioned in Scripture we are apt, very naturally, to picture them with large, bright green leaves of the size

August

and shape of our English oaks; but as this is contrary to fact I will describe the Eastern tree, that we may realise its appearance more accurately.

An acorn, gathered on Mount Tabor, was grown by a friend of mine till the little specimen was old enough to be transplanted into my garden,

PALESTINE OAK.

where it now occupies an honoured place. Its leaves seldom exceed an inch and a half in length, of a dark green, with prickles round the edges.

Unlike our English oaks, which shed their leaves in autumn, these trees are evergreen, and only mark the change of seasons by throwing out pale green shoots in spring. The acorn is small, and has a somewhat prickly cup.

There are three species of oak in Palestine; the one I possess is *Quercus pseudococcifera*, which grows abundantly in Syria. Abraham's Oak near Hebron belongs to this species; it measures twenty-three feet in girth, and the branches are spread over a space ninety feet in diameter. During the severe winter of 1894-5, the weight of snow broke off one of its huge branches, which, when sawn up, furnished sufficient wood to load seven camels.

We owe the ink with which we write to another Syrian oak (*Quercus infectoria*). A small fly punctures its twigs, causing irritation in the flow of the sap, and gall-nuts are formed in consequence. These nuts abound in tannic and gallic acid, and in combination with sulphate of iron and gum they form the constituents of our writing ink.

I have in my museum some of the huge acorn cups of the valonia oak, the third species, known as *Quercus ægilops*; this tree is of great value, as its fruit is much used and largely imported for dyeing purposes as well as ink-making.

We read in Acts xx. 13, that St. Paul, parting from his disciples at Troas, was to meet them again at Assos (to which place they were going by ship); he, 'minding himself to go afoot,' would, in making this journey, pass through groves of valonia oaks, which abound in that part of Asia Minor.

I like to think of the great apostle taking that quiet woodland walk, possibly the last opportunity he ever enjoyed for undisturbed meditation and thought, alone amidst the beauty of nature.

September

'For me, who under kindlier laws belong
To Nature's tuneful quire, this rustling dry
Through leaves yet green, and yon crystalline sky,
Announce a season potent to renew,
'Mid frost and snow, the instinctive joys of song,
And nobler cares than listless summer knew.'

Wordsworth.

September

HEMP (*Cannabis sativa*)

HEMP is such a graceful plant that it would be well worth cultivating in our gardens, if it were only for its rich green leafage; but my hemp bed has now an additional charm, in that its ripe seeds attract all kinds of small birds, which haunt the slender stems, flitting in and out in search of the coveted provender. Four species of titmice are especially active, and may be seen all day long, generally head downwards, creeping about like little mice, twittering cheerfully to each other as they pursue their busy search. These charming birds abound in my old garden, owing to my having attracted them to the place for the last thirty years by keeping a little basket filled with fat outside my dining-room window. This basket has been well known from time immemorial to the successive generations of tits in the neighbourhood.

These birds need a diet of fat in the winter months to enable them to bear the cold, just as the natives in the Arctic regions feed on blubber or whale fat to supply themselves with the warmth needed to sustain life.

It has been a great pleasure to me and to my

friends to watch, not only the titmice, but robins, chaffinches, wrens, and nuthatches enjoying the contents of this basket. Fascinating bird pictures are formed as angry little skirmishes arise, crests are raised and wings outstretched in vehement protest against undue greediness.

Some years ago I placed an empty coco-husk above the basket; it had a small entrance hole at one end, and, as I expected, a blue titmouse built

YOUNG BLUE-TIT.

its nest and laid five eggs in it. The bird was tame enough to allow me to lift down the husk and show to young visitors the little mother sitting on her nest.

When the eggs were hatched the busy parents were hard at work from early morning till late in the evening bringing green caterpillars and small grubs to feed their young family. Knowing that the feeding process began with daylight, and continued without ceasing until dusk, I was able to make a calculation as to the number of caterpillars destroyed by a single pair of these birds in one week, and I found it amounted to about seven thousand six hundred. We can therefore judge of the value of such birds in ridding our gardens of insect pests.

When the young blue-tits were fledged and were leaving the coco-husk one by one to begin life for themselves in the tree branches, I retained one for a little while, that I might take its portrait (as seen in the illustration), and fearing lest it might suffer from hunger, I placed it at intervals in a cage on the lawn, where I had the pleasure of watching the affectionate parents come back to feed it through the bars. The drawing was soon completed, and I need hardly say the little one was allowed to have its liberty.

Bats

'The distant owl
Shouteth a sleepy shout,
And the voiceless bat, more felt than seen,
Is flitting round about.'

Coventry Patmore.

A long-eared bat was brought to me to-day; it had been found in a window-box amongst some flowering plants. When bats will eat either flies or raw meat, one can keep them as pets as long as may be desired, and very curious and interesting they are. This one obstinately refuses food of any kind; I therefore have placed it in some ivy branches, so that when evening comes it may take wing and feed itself.

I remember keeping a similar specimen in my childhood, which became very tame and amusing.

I had more time then to catch flies for it, and it consumed at least forty bluebottles daily. This

LONG-EARED BATS.

fact shows us how useful bats are in keeping down the insect hosts throughout the summer and autumn.

September

Some of the old trees in my woods have hollow stems, and in these bats congregate in large numbers: we are reminded of the fact by the powerful and far from agreeable odour these trees emit as we pass.

The illustration shows a long-eared bat when resting head downwards. A moment or two after alighting it folds up its long ears and places them nearly out of sight under its arms, and then the little creature looks like a mere ball of grey fur.

When on the ground a bat can only scuttle along in a very awkward fashion, as if on hands and knees, and finds great difficulty in taking flight from a level surface.

I have sometimes watched a bat in my room where, on warm summer nights, they occasionally pay me a visit, and I observe that it generally makes its way to a curtain, and climbs up by its hooked wings until it is high enough to dart off into the air.

Bats should never be wantonly destroyed, for they are perfectly harmless and extremely useful. They carry on at night the work that swallows are doing throughout the day—clearing the air of millions of flies, gnats and moths, which would otherwise be a torment to us and very injurious to the farmer and gardener.

Butterflies in Sunshine

I have been fascinated this morning by seeing one of nature's lovely wayside pictures. On the pale lilac flower-heads of a tall sedum six or seven richly-tinted butterflies sat basking in the warm sunshine.

Peacocks, red admirals, and tortoiseshells form a most beautiful mosaic of colour on the soft mauve flowers, and the tints are ever varying as the wings open and shut and reveal the dark, yet brilliant, markings of the under wings.

There must be something very attractive in the honey of this particular stonecrop, for its blossoms are the resort of insects of all kinds; bees, wasps, and flies hover over them all through the hours of sunlight, quite a busy throng ever coming and going.

I like to take a seat for a while near this plant, in order to watch the characteristics of different insects. Some are very business-like, they come for honey only, interfere with nobody, and go away as soon as they are satisfied; others are winged busybodies, buzzing around disturbing peaceable visitors, themselves idle and interfering with those who desire to pursue their own quiet work.

The butterflies are busily engaged, each drawing up nectar with its long proboscis, enjoying sunlight and sweet food. Perhaps they even possess a touch of vanity, and are conscious of some pleasure in exhibiting their lovely wings. If so, they may surely be excused, seeing how truly beautiful they are.

It is well worth while to search amongst nettle leaves in early summer in order to find some of the jet-black, prickly caterpillars from which these handsome butterflies develop, and by keeping them in a box, well supplied with nettle leaves daily, we may see for ourselves the curious chrysalides which shine as if made of gold leaf and hang suspended head downwards, held securely by the small hooks with which the pointed end of the chrysalis is provided. Then in due time out come the exquisite butterflies; their delicate wings, unspoiled by wear and tear, show their bright colours to perfection. Perhaps my greatest pleasure is ultimately to set the captives free, and see them soar away into space to enjoy their brief life of summer flowers and sunshine.

Fungi

We are now being reminded of the approach of autumn by the appearance of various species of toadstools. They spring up in the woods, on our lawns, or on decaying tree-stems; and as the wild flowers and fruits are nearly over, we shall find in them a new and extremely interesting field of study.

Let us collect and examine some of the different kinds of fungi we meet with so abundantly in our daily walks.

Seeing that there are considerably over a thousand species of named fungi, ranging from the microscopic films and moulds which appear on decaying fruit, stale bread and other substances, up to the giant puff-ball, which sometimes measures a foot in diameter, it is clear that we shall only have space for a few general remarks upon the commoner species of fungi we are likely to meet with.

We are all familiar with the edible mushroom (*Agaricus campestris*), so we will select it as a type of the agarics, of which I believe there are several hundred species. It is well to know the right terms to use in describing a fungus, so we will trace the growth of one from its beginning, and learn the parts of which it consists.

If we dig up a mushroom and examine it

carefully, we shall find that it has sprung from a network of white threads, which is called the spawn or *mycelium*. Resting here and there upon this network are small nodules, which will grow into mushrooms in due time; they first appear above the ground as white balls, which, as they rise up and gradually expand, divide into two parts, namely,

AGARICUS CAMPESTRIS.

the cap, which is botanically called the *pileus*, and the stem or stipes.

We notice next a thin membrane which envelops the cap. This is torn away as the pileus enlarges; part of this veil or *volva* remains on the stem, and is called the *annulus*, and part clings to the outer edge of the cap.

We take off the top of the mushroom and reverse it, and now we see thin plates or gills radiating from the centre of the cap. If we cut one of these gills out and lay it on a sheet of white paper, we shall find after a few hours a quantity of dark brown grains, which are called spores. From these arises the mycelium, from which the mushroom springs. These spores vary in colour in different species; some are pure white, some are purple, some have different shades of brown. They also vary in size, but are usually so exceedingly minute

BOLETUS EDULIS.

that one writer declares a single fungus can produce as many as ten million spores.

Thus far my description has applied to the agaricini or gill-bearing fungi, but we will now turn to the second order, the polyporei or spore-bearing fungi.

Under a large tree on my lawn I find this autumn a great abundance of toadstools about the size, shape, and colour of penny buns. If I cared to experiment in that line, I know they would make a perfectly wholesome dish for the table, as they are the well-known *Boletus edulis*.

The pileus is a rich shining brown colour above, but when we examine it beneath we shall see, instead of the gills of the agaric order, an orange-coloured spongy substance consisting of tubes or spores.

Other species of this order are woody excrescences growing out of decaying tree-stems. A material

HYDNUM REPANDUM.

called amadou, used for making fusees, is obtained from several kinds of *polyporus*. Yet another species is *Merulius lacrymans*, so well known by the name of 'dry rot,' which is far too frequently met with in old timbered houses.

As the threads of the mycelium penetrate the wood, they reduce it at last to a state of absolute rottenness. This process may go on quite secretly for years, but suspicious cracks become apparent in our wainscot, and when some panels are removed there we see visible evidence of dry rot. Large patches of a grey velvety substance are spreading everywhere, covered with drops of water which gives the specific name of *lacrymans* (weeping) to this most destructive fungus.

In the third order of fungi we find beneath the pileus spiny projections or teeth. If we happen to light upon *Hydnum repandum*, a species not uncommon in woods and damp shady places, we can observe in it a good specimen of this structure. Then again we notice the curious *Clavarias*, mauve-coloured, white, yellow and bluish grey, which spring up on our lawns at this season. They dry very readily, and form interesting subjects for a collection.

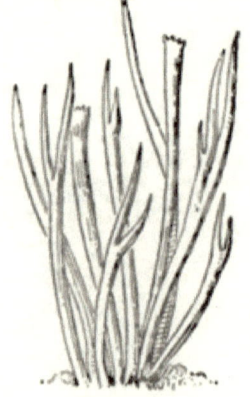

CLAVARIA.

Pezizas are also worth searching for. I found a brilliant orange-coloured one on our common to-day, and could not resist bringing it in, so as to watch it giving out its spores when breathed upon. They are shot out like little jets of smoke, and it is amusing to see the fungus thus energetically sowing itself far and wide.

Any of my readers who may desire further information on this subject will find in Dr. M. C. Cooke's *British Fungi* an excellent guide into this field of special study to which the specimens of to-day have drawn our attention.

ACANTHUS

> 'And hear the emerald-coloured waters falling
> Through many a woven acanthus-wreath divine.'
>
> *Tennyson.*

I always look with interest at this handsome plant, with its finely shaped, glossy green leaves. It is not merely its beauty as a foliage plant that attracts me, but I am reminded of the many classic associations which have clustered around it.

ORIGIN OF CORINTHIAN ORDER.

It is said that the graceful form of these acanthus leaves and their mode of growth suggested to Callimachus, a sculptor who lived nearly four hundred years before Christ, the first idea for the decoration of the capital of the Corinthian column. It is easy to suppose that some vase or basket accidentally overgrown by this plant would furnish an excellent model to an observant eye; and skilful hands would soon adapt the curved leaves into the sculptured decoration we so much admire in buildings ancient and modern.

Acanthus leaves appear in the dresses of the figures

ACANTHUS MOLLIS.

on Etruscan vases, and they were often cut out in purple cloth, and formed into richly embroidered borders for Roman garments.

I have given a drawing of this plant, since it may

be of interest to my readers, when they pay a visit to any museum of antiquities and sculpture, to endeavour to trace the many ways in which these beautiful leaves have been used for decorative purposes.

BROOM-RAPE (*Orobanche speciosa*)

About one hundred species of these parasitic plants appear to be known. Strange uncanny growths, deriving their nourishment as they do from the roots of other plants, we may fairly regard them as the thieves and vagabonds of the vegetable kingdom. It is not difficult to find certain of our English species.

The lesser broom-rape grows abundantly in clover fields and on gravelly heaths; the tall brownish spikes of the *Orobanche major* can readily be discovered growing on various plants, such as broom, furze and other species which bear pea-shaped flowers.

Some seeds of broom-rape from Southern Europe having been sent to my gardener, he tried the experiment of sowing them in such a manner that I might watch their growth from their early stages on to maturity. Some broad beans were sown, destined to be the victims of the parasite, and when they had germinated the seeds of the broom-rape were carefully introduced below the surface of the soil. In due time there appeared about a dozen spikes of the *Orobanche*, clustering round the broad bean plant, which was then two feet high and already bearing pods.

The seeds of the parasite must have germinated and in some mysterious way discovered the presence of the roots upon which it was their nature to grow.

September

To the bean root the broom-rape seedling adheres, and its stem swells into a bulb, as shown in the

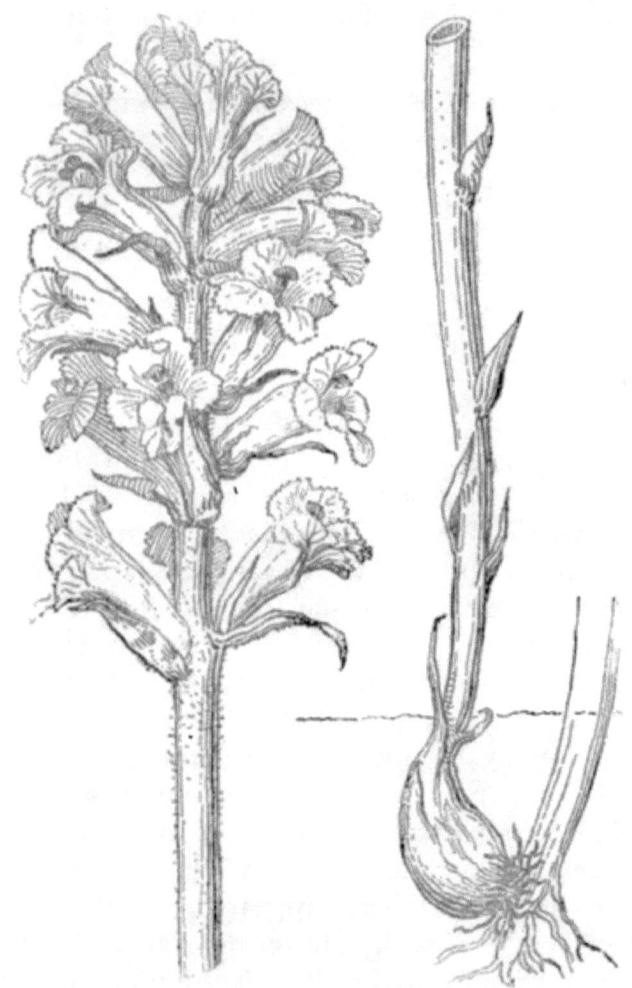

OROBANCHE SPECIOSA.

drawing. Now, having established itself, it has only to continue sucking nourishment from its host, growing and flourishing, and in the case of this foreign species sending up really handsome spikes of lilac-tinted flowers.

The Rev. Professor G. Henslow, speaking of this broom-rape, says,[1] 'A field of beans just outside Cairo looked at a distance like some nursery ground for gorgeously flowering herbaceous plants in masses, as there was more of the broom-rape to be seen than beans. It consisted of tall spikes some four feet in height, densely covered with white, yellow, and lavender-coloured blossoms of different shades.

'It would make a splendid herbaceous border plant, of course associated with some broad bean plants for it to live upon. By dumb show I pointed out to an Arab the necessity of cutting them down, pointing to some dead bean plants. He only shrugged his shoulders, smiled and said, "Kismet," and then walked away.'

I confess it was a little sad to watch my bean plant, thus preyed upon, diminishing in vigour and at last dying a victim to scientific experiment. I could not but speculate as to the use and intention of the creation of these parasitic plants. I have arrived at no satisfactory conclusion, and must leave it as a problem for my readers to solve.

[1] *The Gardener's Chronicle*, July 30. 1898.

October

'Where are the songs of Spring? Ay, where are they?
Think not of them, thou hast thy music too,
While barred clouds bloom the soft-dying day,
And touch the stubble-plains with rosy hue;
Then in a wailful choir the small gnats mourn
Among the river sallows, borne aloft
Or sinking as the light wind lives or dies;
And full-grown lambs loud bleat from hilly bourn;
Hedge-crickets sing; and now with treble soft
The redbreast whistles from a garden-croft,
And gathering swallows twitter in the skies.'

Keats.

October

Papyrus

AN object of much pleasure to me in my hot-house is a fine specimen of the Egyptian papyrus reed, with stems fully eight feet high, growing with remarkable luxuriance and beauty. I never look upon its graceful flowering plumes without being reminded of a chain of interesting associations.

The infant Moses was laid amongst these so-called 'bulrushes,' which then grew along the margin of the Nile. The 'ark' in which the child was laid was formed of papyrus stems, and the small cradle would be readily concealed amongst 'the flags by the river's brink.'

This plant is now wholly extinct in Egypt, although it still grows abundantly in the marshes of the White Nile in Nubia. A verse in Isaiah, in the revised edition (chapter xviii. 2) shows that in ancient days even boats were made of papyrus, and a modern traveller speaks of the plant being still used by the Abyssinians for the same purpose.

It has lately been discovered that mummy cases were sometimes constructed of old papyrus rolls, and many very ancient and valuable writings have been obtained by soaking these coffins in water until, with the exercise of great patience and care, the original strips of papyrus could be separated and then pieced together, so that the writing can be deciphered.

Many years ago I listened to an address by Mr.

PAPYRUS SYRIACUS.

John MacGregor, in which he gave a vivid description of his explorations in Palestine, and mentioned his discovery of an immense extent of papyrus growing in the upper reaches of the Jordan. The snow melting from Mount Hermon trickles down in small streams, forming marshes five miles in length and about three miles broad, closely filled with papyrus stems from eleven to fourteen feet in height. Such a reedy swamp would have been impenetrable but for a narrow channel winding through it, which enabled Mr. MacGregor to make his way with the Rob Roy canoe until he reached the open waters of Lake Merom.

After examining some fragments of ancient papyri, I felt sure that it would not be impossible to make paper of the same kind from my own specimen, and this was the way in which I succeeded in the manufacture. I cut a stem eight feet long into lengths of about six inches, and with a sharp knife sliced off the green rind from its three sides, and cut the remaining white pith into very thin layers.

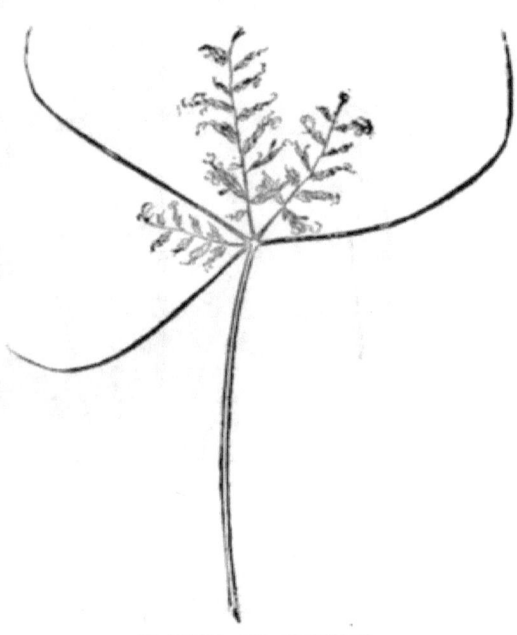

FLOWER OF PAPYRUS.

Having a hot iron ready at hand, I quickly laid the strips of pith side by side, each a little overlapping the other, on a sheet of white paper, and when it was covered I placed another layer upon it at right angles to the first layer. With a sheet of paper to keep the iron from adhering to the pith, I pressed the two thicknesses of pith firmly together until they were closely united. In about a quarter of an hour, by repeated ironing, I found I had made a piece of light grey material exactly resembling the ancient papyrus which was

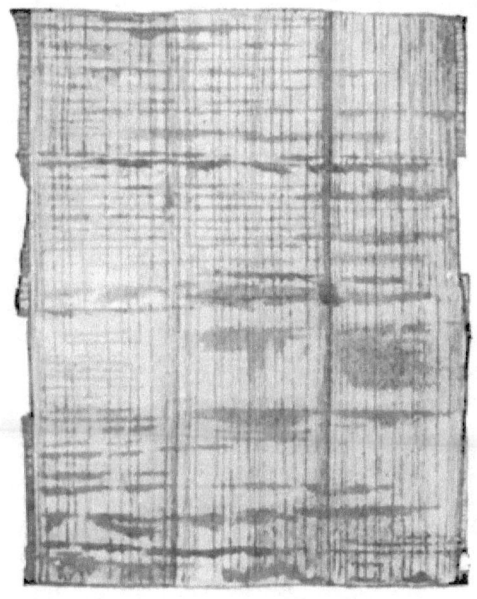

PAPYRUS PAPER.

my pattern. The sap of the plant appears to possess an adhesive quality, so that no gum is required, the action of heat being sufficient to make the strips unite into a flat even surface, suitable to be written upon with a quill and ordinary ink.

In olden times the young succulent shoots of

this reed appear to have been used as an article of diet, and when stewed and served with a rich kind of sauce it was reckoned, by both Jews and Egyptians, as a table delicacy.

As I have already remarked, the chief interest which centres in this plant is the fact of its great antiquity. In the British Museum we may see papyrus rolls which were inscribed three thousand years ago. The key to the ancient languages has been discovered, and the learned in such matters can decipher that which was penned in the days when the Israelites were toiling in Egypt, and many deeply-interesting facts concerning Scripture history have in this way been brought to light.

GALLS

> 'The flowery leaf
> Wants not its soft inhabitants. Secure
> Within its winding citadel, the stone
> Holds multitudes. But chief the forest boughs,
> That dance, unnumbered to the playful breeze,
> The downy orchard, and the melting pulp
> Of mellow fruit, the nameless nations feed
> Of evanescent insects.'

The pretty wild-rose gall, popularly known as Robin's pincushion, or Bedeguar gall (*Rhodites rosæ*), shows itself very conspicuously in the hedges at this season.

It is like a bunch of finely divided green moss-sprays, brightly tinged with crimson, and is produced by a small four-winged fly, *Cynips rosæ*.

Early in June this glossy black fly lays its eggs in young briar-shoots; the presence of these eggs interrupts the flow of the sap, and woody tissue begins to form around the eggs.

WILD-ROSE GALL.

If we take a gall of this kind in an early stage of growth and cut it in half, we shall find several little cells, each containing a small white grub. These larvæ continue to grow to their full size, and then remain quiescent until the following spring, when they change to chrysalides. The perfect fly emerges when the days become warm and sunny.

The oak tree is victimised by gall-flies innumerable. They lay their eggs in its leaves, branches, flowers, and roots, no part of the tree being exempt from their attacks.

OAK-LEAF GALLS.

Mr. Stephens, a great authority upon insects, says that there are nearly two thousand species of insects which prey upon the oak tree, either as gall-flies depositing their eggs in its substance, or as caterpillars feeding upon its leaves.

OAK FLOWERS—GALLS
(*Natural size*).

A collection of oak-galls would therefore show a great variety of forms, and might profitably occupy our attention this autumn.

I have been picking up leaves entirely covered with bright crimson spangle-galls. Such leaves lie on the ground all through the winter whilst the grubs are maturing; and if we find some of these leaves about the end of February, and keep them in a bottle, slightly moistening them from day to day,

TURKEY OAK.

the flies will hatch, and we can see for ourselves *Cynips longipennis*, the exact species that has caused the spangle-gall.

The large round gall shown in the illustration is the product of a species of *Cynips*, and is beautifully coloured with a pinkish crimson on one side.

October

The smaller galls on the oak flowers are in an early stage; they grow to the size of red currants, and then drop to the ground, the flies hatching out of them in the following spring.

English Oaks

The oak foliage has now turned a soft golden brown, which sheds a kind of sunlight glow over

COMMON ENGLISH OAK.

the landscape. The squirrels are extremely busy collecting and storing acorns for their winter food; and so carefully do they secure not only acorns but nuts and beechmast, that in a week or two it will be almost impossible to find any woodland fruits beneath the trees.

This is the best season of the year to study our native oaks, because we can easily identify them by their acorns. We possess in reality but one indigenous species, known as *Quercus robur*, but there are two varieties, *Quercus pedunculata*, which

LONG-STALKED OAK.

has acorns on much longer stalks than *Quercus robur*, and *Quercus sessiliflora*, which produces its acorns clustered together upon the twigs without any stalks. Its leaves are also broader and more closely grouped together.

The deeply-cut leaves of the imported Turkey

oak, *Quercus cerris*, and its charmingly mossy-cupped acorns readily distinguish it from our English species.

A tree of this kind stands on my lawn, and every autumn, for some years past, on a special day, when

SESSILE OAK.

the rooks by instinct have found out that the fruit is ripe, they come from my rookery in flocks to feast upon the acorns and carry them away, as I believe, to some hiding-places of their own.

All day long the great birds are winging their way to and fro, cawing and rejoicing over the spoil, until they leave the tree entirely stripped, with only a carpet of empty acorn-cups strewing the ground beneath.

In times of scarcity we should do well to imitate the squirrels and store up our acorn crop, for when dried, roasted, and ground into flour a not unpalatable kind of coffee can be made of acorn kernels. I can speak from experience, for some years ago I had this coffee made, and used it as a tonic beverage. I cannot say it had the aroma or flavour of true coffee, but it made a fair substitute for it, and it is believed to be wholesome and strengthening.

The Cedar of Lebanon (*Cedrus Libani*)

Towards the close of this month I always find my great cedars covered with their cone-shaped male catkins. I see now that they are just ready to shed clouds of pollen; but, plentiful as these blossoms are, it is the rarest thing to be able to discover any but male catkins; the female ones appear almost invariably to grow upon the upper branches, where they are quite inaccessible.

For fifteen years I carefully watched for these small cones, wishing to observe them in their early stage, but failed to find a specimen until a few years ago, when one of my cedars obligingly produced some fruit on the lower branches. The drawing will show my readers the two kinds of blossom. The yellow pollen-bearing catkins drop off in a few weeks, whilst the fertilised cones remain, and gradually increase in size until they are easily to

October

be discerned upon the branches, and are of an exquisite pale tint like shaded sea-green velvet.

Cedar catkins are fertilised only by the wind, which carries the pollen from one blossom to the other. The buoyancy of the pollen-grains is much aided by two little bladders with which each grain is furnished, and which can be easily seen by the aid of a microscope.

The cones are borne on the upper side of the horizontal branches, and are not fully ripe until the autumn of the third year. They do not then fall off like other fir-cones, but the scales and seeds become loosened, and drop to the ground.

CEDAR CATKINS.

Of these grand mountain trees there are three species, *Deodar* of the Himalayas, the *Cedrus atlantica* of the Atlas range in North Africa, and the cedar of Scripture, of which, besides many smaller ones, twelve patriarchal specimens may still be seen on Mount Lebanon. These grow at an elevation of about 6000 feet above the sea, their trunks measuring from forty to forty-seven feet in circumference at the base.

It is said that many years ago a Frenchman, who was travelling in the Holy Land, found a little seedling among the cedars of Lebanon, which he wished to bring away as a memorial of his travels. He took it up carefully, and for want of a better flower-pot he planted it in his hat, where he kept and tended it. The voyage was stormy and tedious, so that the supply of fresh water fell short, and only half a glass a day could be spared for each traveller. The little tree was allowed its share of even this scanty allowance, and although the traveller suffered from his self-denial, the little tree flourished, and had attained the height of six inches when the vessel arrived in port.

At the custom-house the officers thought the hat must surely contain some valuables on which duty ought to be paid, and it needed much earnest pleading on the part of the traveller to induce them to spare the cherished seedling.

Eventually it was allowed to pass through unharmed. It was then taken to Paris, and found a place in the Jardin des Plantes. In the course of years it grew into a noble tree. It lived on for over a century, until, sad to relate, the beautiful tree had to be cut down to make way for a railroad.

It would be quite possible to grow our own cedars with the exercise of patience. A seed I planted out of a cone from Lebanon remained dormant for twelve months in the earth before the young plant made its appearance. Probably if the seeds were soaked in water for a few days before they are planted, it might tend to hasten the process of germination.

The Liberation of Seeds

The capsules of the cyclamen are now opening; they are curiously spotted inside, and look like small brown flowers. The twisted stem is coiled around the capsule, and keeps it closed until the seed is perfectly ripe. Then it uncoils, the segments curl backwards, and the seeds are allowed to drop out and sow themselves.

The iris, the datura, and a large number of other plants produce capsules which open their valves when ripe and allow their seeds to escape, and this is perhaps the simplest mode of liberating ripe seed;

YELLOW IRIS CAPSULES.

but at this season, when so many plants are producing their fruit, we shall find it quite interesting to note some of the many other curious modes by which seeds are dispersed.

We observed in the spring the fruits of the sycamore, maple, and hornbeam, which are furnished with a *samara*, or thin membrane, so that the autumn breezes may bear them flying through the air, and sow them far away from the parent tree.

The wild balsam affords a good example of

dispersion by elastic force. The valves curl up and jerk the seeds in all directions.

Heartsease, woodsorrel, wild geranium, and many other plants scatter their seeds in the same manner.

The wild pimpernel has a special way of sowing itself by dropping half of its cup-shaped capsule. This, being a common field flower, can easily be found and examined.

Almost all such flowers as the dandelion, goatsbeard, succory, belonging to the extensive order of Compositæ, have seeds more or less feathered, so that

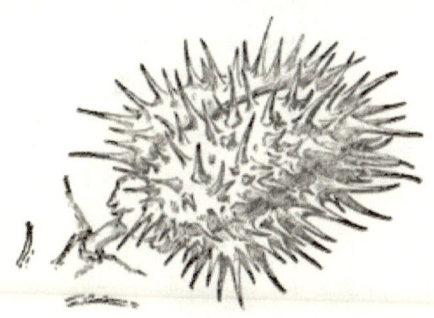

DATURA.

they may be wind-dispersed; but, being so common, I need not describe these in detail.

A tree which grows on a mountain in the Cape Colony is known as the silver tree (*Leucadendron argenteum*), from its leaves and cone being so thickly covered with shining white hairs that they look as if they were made of silver. The leaves hang vertically, exposing only their edges to the sun; consequently the trees afford but little shade, only a criss-cross of fine lines of shadow is thrown upon the ground. I mention this tree because its cone produces a remarkable kind of seed. Reference to

October

the plate will show the four feathery plumes by which the wind wafts the seed through the air. They rise out of the dry capsule, and from it the heavy seed hangs at the end of a slender thread, the whole arrangement being somewhat like a small parachute. The silvery cone is a beautiful object in itself, and when fully ripe, one of these curious seeds emerges from under each of the overlapping scales

The capsules of the poppy, campanula, and snapdragon allow their seeds to escape through small pores which, being highly sensitive to dryness and

COLUMBINE.

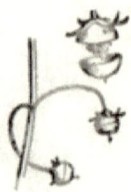

PIMPERNEL.

moisture, open and shut according to the changes in the weather. We can easily observe these small trapdoors under the upper rim of the poppy-head, and in the other plants I have mentioned the openings are in the upper part of each segment of the capsule. The columbine has a five-pouched seedpod opening at one end when ripe, and bending down to sow its contents.

Space will not allow me to notice the many other modes by which plants perpetuate their species, some by hooked seeds which cling to passing animals, some, like the cotton-grass, by very long silky hairs. Others, and perhaps the most curious of all, are those highly sensitive to moisture and dryness, which by expanding and contracting are enabled to creep

SILVER-TREE SEED.

CYCLAMEN CAPSULES.

along the ground. All these will afford pleasant hours of study to those who like to investigate nature's secrets, as seen in the commonest things which lie about our daily path.

November

'The pale descending year, yet pleasing still,
A gentler mood inspires; for now the leaf
Incessant rustles from the mournful grove,
Oft startling such as, studious, walk below,
And slowly circles through the waving air.
Fled is the blasted verdure of the fields;
And shrunk into their beds, the flowery race
Their sunny robes resign.'

Thomson.

November

BIRDS' FEET

IT is quite worth while to observe the characteristic variations of form in the feet of birds; they will be found to be wonderfully adapted to the kind of life they live and to the food they have to subsist upon.

When I chance to find a dead bird, I usually retain its feet, attaching them to a card and allowing them to dry slowly within the fender. In this way I have made a small collection, which includes specimens from the various divisions of the bird kingdom, and very useful I often find it for purposes of reference.

Eagles, hawks and owls (all of which are known as raptorial or seizing-birds) are provided with strong, sharp claws, with which they clutch and kill the animals and birds they feed upon.

A glance at the claws of the owl shows us that the grip of such a foot cannot fail to squeeze to death a small rat or mouse. An owl swallows a mouse whole, and next day the bones and fur are thrown up in the form of a small grey pellet; the amazing number of bones to be found in these pellets goes far to prove the value of owls as rat and mouse destroyers.

The water-rail and moorhen appear to be links between land and water birds; they can swim short distances by means of a membrane on each side of the toes, but their lives are mostly spent in threading their way through the sedgy herbage which grows on the margin of ponds and lakes.

The true swimmers (*natatores*) include all such birds as swans, geese and ducks; we can see how easily they propel themselves in any direction by means of their webbed feet, which so admirably fit them for their aquatic life.

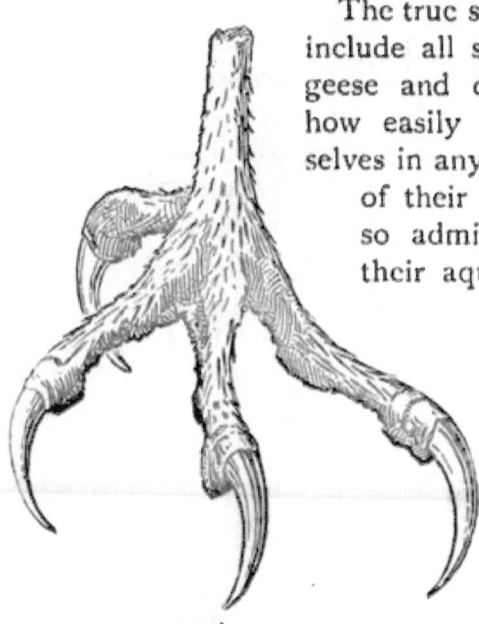

OWL'S FOOT.

I possess the feet of a curious foreign bird, the jacana, which frequents Brazilian lakes, where water-lilies abound, and by means of its long toes it can walk upon the leaves and find its insect diet.

The light weight of the bird is spread over a considerable area, so that it is borne up on the leaf-covered surface of the water much in the same way that a traveller in Arctic regions is supported on his journeys by means of his wide-spreading snow-shoes.

A ptarmigan affords us a specimen of a bird well protected against the effects of cold by having its feet thickly furred to the very claws. Its plumage is pure white in winter, so as not to be readily seen upon the snowy ground, but in summer the feathers change to grey and brown, colours which make the bird inconspicuous amongst grey rocks and heather.

Some trumpeter pigeons I kept at one time had oddly feathered feet; one could not imagine for what purpose the feathers grew along the toes; they seemed neither useful nor ornamental; I came to the conclusion that they must be a freak of nature, and one of the results of domestication.

Grain-eating birds (*gallinaceous*), such as turkeys, fowls, pheasants, and a large number of other species, are provided with very strong feet armed with horny toe-nails, to enable them to scratch up the earth in order to find their food. The foot of a common fowl will afford us an example of this class of birds.

JACANA'S FOOT.

I would call attention to the long claw of the lark, the use of which was, I believe, a puzzle to

SNOW-SHOE.

naturalists, until it was discovered that by its means

FOOT OF PTARMIGAN.

the bird was enabled to grasp and carry away its eggs when any danger threatened their safety.

November

The lark's nest being built upon the ground is

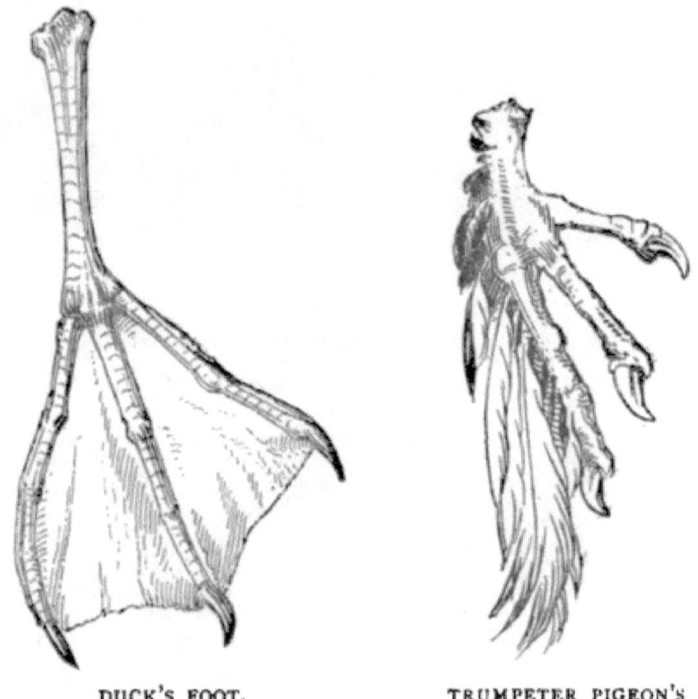

DUCK'S FOOT.

TRUMPETER PIGEON'S FOOT.

exposed to many dangers, and when the mower's scythe has laid it bare, the mother-bird has been observed carrying away the eggs one by one in her long-clawed foot. It has also been suggested that the hind claw may tend to break the shock of alighting on the

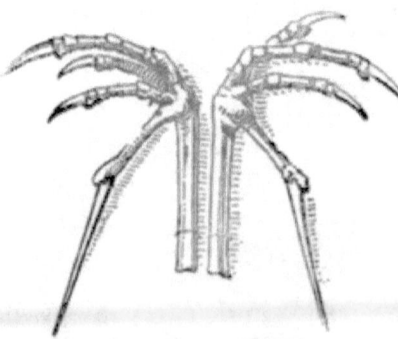

FEET OF LARK SHOWING LONG HIND CLAW.

ground from a great height; in either case it offers an interesting instance of provision for a bird's special need.

When we reflect that there are more than ten thousand species of birds, inhabiting every variety of situation and fitted to every climate, we may form some idea of the need of adaptation in their structure.

In these slight remarks on birds' feet, I only attempt to draw my readers' attention to a very wide subject, which they may like to study further from time to time as opportunity may occur.

Flies Killed by Fungus

A very miserable fate is now overtaking some of our common house-flies. If they happen to come in contact with a very minute fungus known as *Empusa musci*, one of the spores throws out a tube and penetrates the body of the fly, where it will grow and multiply its cells until it has gradually eaten out the interior of the insect.

I found a specimen of one of these victims on the window-pane to-day. The fly's body was swollen and fixed to the glass; the wretched insect was dead, the fungus was showing on the outside of its body, and all around it the white spores lay like a misty halo upon the glass.

FLY KILLED BY FUNGUS.

The fungus has the power of throwing its spores some little distance off, and if one of them falls upon a living fly the same process is again repeated, and before long the victim dies this miserable death.

The caterpillar of the common white butterfly is frequently attacked, and dies in the same way when seized upon by a species of minute fungus.

The Nautilus

The shell of the common nautilus, when divided lengthways, affords a beautiful example of delicate structure. It is the dwelling of a species of cuttlefish found in the Indian Ocean.

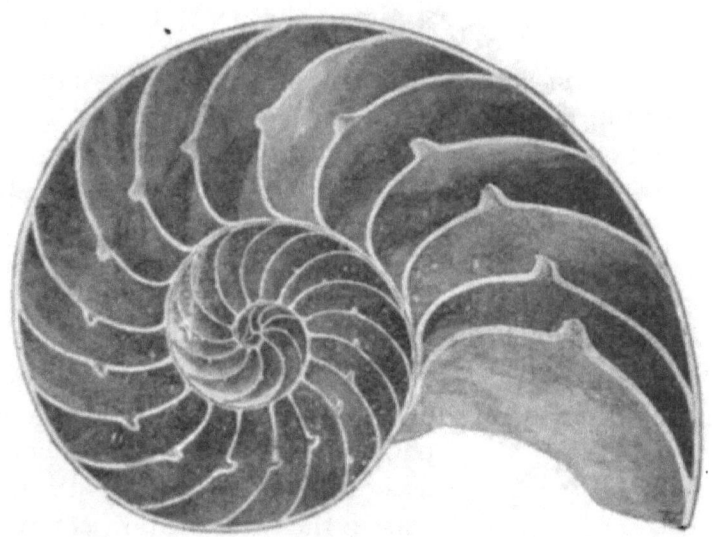

CHAMBERED NAUTILUS.

The creature lives only in the upper compartment of its shell, whilst below it are thirty-six exquisitely graduated air-chambers lined with mother-of-pearl.

This cuttle-fish has numerous tentacles or feelers, on which it sometimes crawls like a snail at the bottom of the sea. It is a deep-sea dweller, but at times it rises to the surface, and swims through the water by drawing in air and then violently ejecting it, thus progressing backwards by a series of jerks. The shell is as hard and smooth as porcelain, and is marked outside by a series of dark brown wavy lines.

Dr. Oliver Wendell Holmes, in his beautiful poem *The Chambered Nautilus*, draws a delightful lesson from the formation of this shell.

> 'Build thee more stately mansions, O my soul,
>   As the swift seasons roll,
>   Leave thy low-vaulted past!
> Let each new temple, nobler than the last,
> Shut thee from heaven with a dome more vast,
>   Till thou at length art free,
> Leaving thine outgrown shell by life's unresting sea.'

Another species, known as the paper nautilus, has a pure white and exquisitely fragile exterior, in form resembling the common nautilus, but without any chambers inside. Indeed, instead of being a solid and polished substance, its shell is of an extremely delicate and thin material, furrowed into long wavy wrinkles.

For ages this shell has been represented, as in the accompanying drawing, sailing along on the surface of the sea like a fairy bark, with two tiny sails uplifted to catch the wind. It was said to have given to man the idea of navigating the ocean; Aristotle thus described it. Pope writes, 'Learn of the nautilus to sail.' Montgomery and other poets allude to its being seen thus floating on the sea; but, alas, the rude hand of science has brushed away the charming poetic fancy, and we are told that the two flattened membranes, which were supposed to be sails, are only used for the prosaic

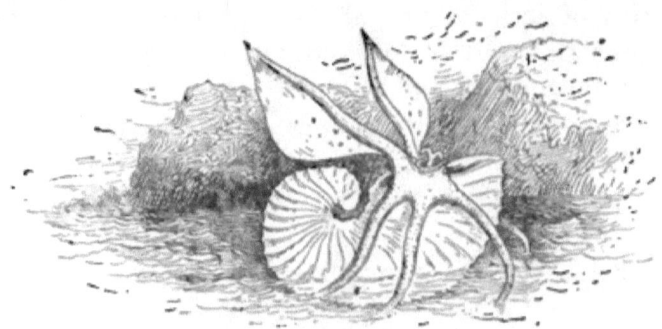

ARGONAUT, OR PAPER NAUTILUS
(*Mythical*).

purpose of secreting calcareous matter, in order to repair the shell when injured.

We do not readily part with such a charming vision as the poet thus describes:—

> 'Light as a flake of foam upon the wind,
> Keel upwards, from the deep emerged a shell
> Shaped like the moon ere half her orb is filled.
> Fraught with young life, it righted as it rose,
> And moved at will along the yielding water.
> The native pilot of this little bark
> Put out a tier of oars on either side;
> Spread to the wafting breeze a two-fold sail,
> And mounted up, and glided down the billow,
> In happy freedom, pleased to feel the air,
> And wander in the luxury of light.'
>
> *Pelican Island*, by MONTGOMERY.

One could wish to be a fairy watching this little skiff come towards one across a halcyon sea!

Seed Mimicry

It is rather difficult to imagine for what purpose seeds of certain plants have been created with such a strong resemblance to insects, shells, twigs, and other objects. In the drawing, a land shell (*Helix Lapicida*) is shown together with the seed of *Medicago*

HELIX LAPICIDA AND MEDICAGO HELIX.

CATERPILLAR SEED
(*Natural size*).

helix. It will be seen that the one is a counterpart of the other. The curly seed of another species of *Medicago* instantly reminds one of the large green caterpillars which abound in cabbage plants, and whose habit it is to curl up the moment they are touched.

The seeds of many plants strongly resemble beetles, and others are like hairy spiders. We have already noticed the bird's-foot trefoil as one of these mimicking plants, and many others might be mentioned. It may be that birds, deceived by the appearance of these seeds, take them for insects, and finding they have been mistaken drop the seeds at a distance from the parent plant, and thus ensure their dispersal into fresh soil and surroundings.

Hidden Lives

As we pursue our nature studies, we cannot fail to be struck by the fact that there exists all around us curious hidden lives of creatures unknown to us. These are only revealed when, by chance, our attention is called to some trace left behind them which excites our curiosity. Then, indeed, investigation may often lead to interesting discoveries.

We may frequently find snail shells on hedgebanks with no living snail inhabitant, but half filled with dry clay. Any one might suppose these had become thus filled by accident, but if we take the shells home and place them in a box with muslin or net over the top, we shall find that specimens of the mason bee (*Osmia aurulenta*) will in due time hatch out of the mud-cells in the snail shell. It is the habit of this bee to choose an empty shell as a cradle for her young. The *Osmia* collects little pellets of mud, and with it she forms cells to contain her eggs and food for the grubs which will hatch out of them.

NEST OF BEE IN SNAIL SHELL
(*Natural size*).

The mother-bee carries out this arrangement in summer, and leaves her nursery to itself; winter passes by, and in the following spring the young bees emerge from the snail shell to begin life on their own account. Their instinct teaches them to do exactly as their unseen parent did, and so they perpetuate their species in a similar manner.

In the angles of brickwork we may often see a small mass of grey mud, which looks as if it might have been thrown there by a passer-by. We have only to investigate with a penknife and remove a portion of the mud wall, and we shall find there also is a hidden life history.

A small species of wasp forms its cells in the angle, and covers them with grey mud, which hardens and protects her eggs through the frosts of winter, so that in the coming summer her young will come out in safety and begin their life work.

I was much interested this year in a dwelling-place which happened to be new to me, and may possibly be so also to my readers. A minute fragment of a leaf was swinging at the end of an invisible thread depending from an oak branch. Something led me to examine it, and I found it was a cone-shaped dwelling inhabited by a lively little caterpillar. I could hardly believe my eyes, so minute was the whole thing, and yet so perfect. Evidently the tent-dweller knew what he was about, and was carrying out his life-destiny in thus descending from the oak tree to the ground.

The case was about the size of the capital letter I on this page; it was formed of atoms of oak leaf glued together. The caterpillar in his house careered about on the surface of an oak leaf where I had placed him, and when he grew a little fatigued by his journeying, he came to anchor for the night by fixing his tent in some way firmly to the surface of the leaf. I believe this small specimen was the larva of one of the very minute species of moths called *Tineas*.

We can hardly fail to notice that some of the

LEAF MINED BY MOTH LARVÆ.

leaves of such shrubs as honeysuckle, bramble snowberry, and other species have curious intricate patterns traced upon them; and we may have speculated as to the cause of this. Let us gather a bramble leaf such as that shown in the drawing, and we shall be able to trace a life-history begun and ended in that one leaf.

A very minute moth lays its eggs in the leaf in early summer, and the grub which comes out of it mines its way between the fibre and the outer skin of the leaf, feeding, as it proceeds, upon the green substance it finds there. The mark that it makes is at first like a fine white thread, but as the larva grows its tunnel increases in size until the grub is full-grown; then it emerges and falls to the ground, changes into a chrysalis in the earth, and remains there until the following summer, when the moth hatches, and the life-history begins over again. In the drawing there are the tunnels of two of these leaf-miners shown, and the gradual increase in size can easily be traced.

There are, I believe, many hundred species of these exceedingly minute moths, each one choosing some special plant in which to deposit its eggs. The perfect insects are in some cases extremely beautiful, like little jewels adorned with gold and silver fringes.

I have chosen these few specimens of 'hidden lives' simply to stimulate observation; similar cases exist in every department of nature, and will amply repay careful study.

December

'A winter such as when birds die
In the deep forests, and the fishes lie
Stiffened in the translucent ice, which makes
Even the mud and slime of the warm lakes
A wrinkled clod, as hard as brick; and when
Among their children, comfortable men
Gather about great fires, and yet feel cold;
Alas! then, for the homeless beggar old.'

Shelley.

December

The Holly

THE hollies are reflecting the bright morning sunshine which glistens on their polished leaves.

These are, as far as I know, the only trees which have sharply spiked leaves on the lower branches only, to defend the foliage from the attacks of browsing cattle. Higher up out of reach, the leaves are perfectly smooth and unarmed, resembling those of the camellia. It is difficult to believe such differing leaves can belong to the same tree.

Southey's well-known lines refer to this peculiarity in the holly leaves.

'O reader! hast thou ever stood to see
  The Holly Tree?
The eye that contemplates it well perceives
  Its glossy leaves,
Ordered by an Intelligence so wise
As might confound the Atheist's sophistries.

Below a circling fence, its leaves are seen
  Wrinkled and keen;
No grazing cattle through their prickly round
  Can reach to wound;
But, as they grow where nothing is to fear,
Smooth and unarmed the pointed leaves appear.

I love to view these things with curious eye,
    And moralise;
And in this wisdom of the Holly Tree
    Can emblems see,
Wherewith perchance to make a pleasant rhyme,
One which may profit in the after-time.

THE HOLLY.

Thus, though abroad perchance I might appear
    Harsh and austere;
To those who on my leisure would intrude
    Reserved and rude:
Gentle at home amid my friends I'd be,
Like the high leaves upon the Holly Tree.'

OTOLITHS

As the ground is frozen, and all nature seems asleep at this wintry season, we must defer our out-of-door rambles and go into my museum for some otoliths for to-day's study.

The word sounds like something very scientific and out of the way, and yet, without knowing it, these objects have been constantly upon our dinner-plates, for they are little snow-white bones to be found in the heads of haddock, whiting, gurnard, and cod. How these little stone-like bodies assist the hearing of fishes is, I believe, not very clearly known, but that is supposed to be their use in the economy of the fish.

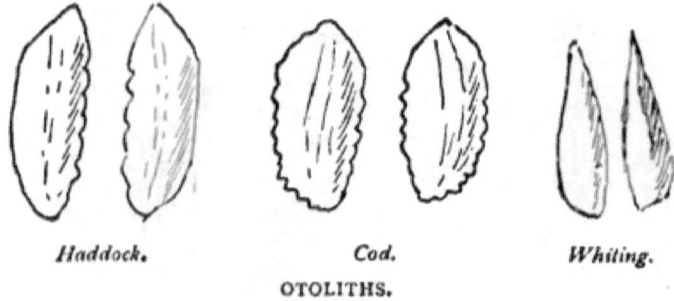

Haddock. Cod. Whiting.
OTOLITHS.

One exists in each lobe of the brain, so that if we wish to find them we must completely divide the head of a whiting, when boiled, and there hidden on either side we shall discover the otolith. It appears to be quite unattached to the skull, and simply lies in its cavity to aid in the conveyance of sound to the fish's brain.

I may mention a use to which I have put these ear-bones with a rather good result.

Having a store of rose-beetle wings and otoliths, I resolved to decorate a banner screen with them in this fashion. I traced, on a piece of rich dark green satin, a flowing design of jasmine sprays.

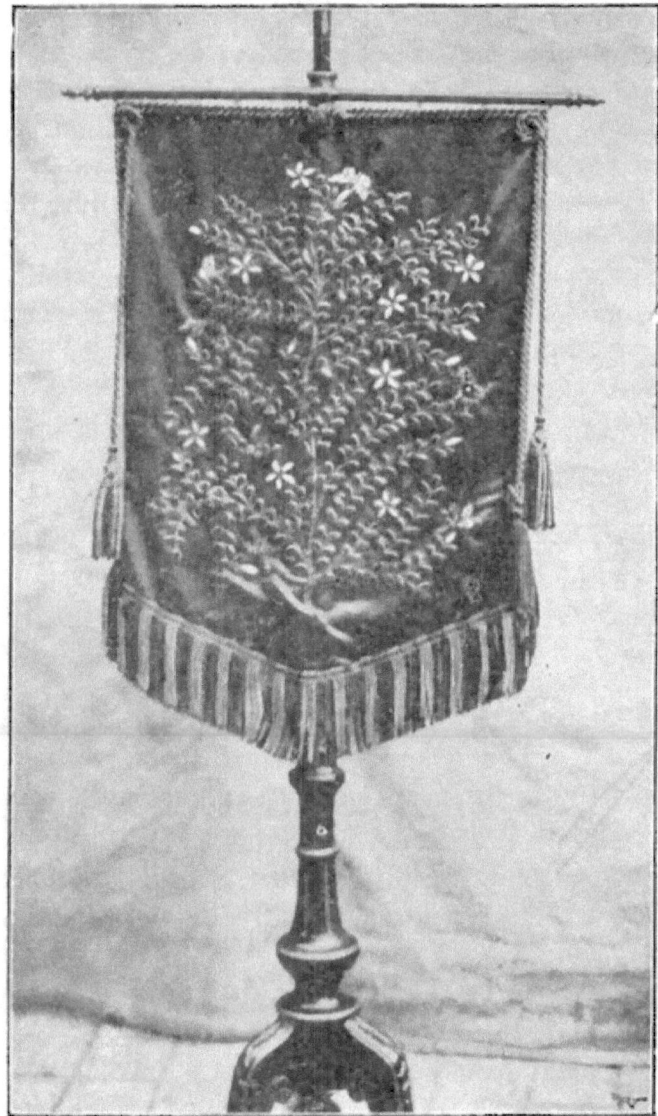

AN OTOLITH SCREEN.

With fine white silk I tacked on sets of five otoliths starwise, each star to represent a jasmine flower, while the beetle wings did duty for the leaves. Each otolith and beetle-wing was edged round with fine gold braid, which kept them firmly in place, and also formed the connecting stems. The beetle wings had to be pierced with a very small needle and each sewn on separately with fine green silk. The ear-bones will not admit of piercing, so two stitches of white silk across either end attached them firmly to the satin.

The plate will give an idea of the effect, which is remarkably good, and my rather original banner screen has, I must say, been much admired.

Rose-beetles are not usually to be found in any number, but an even better result may be obtained by Indian beetle-wings, which are sold at all Berlin-wool shops.

The otoliths must of course be saved up from our daily repast until we have sufficient for the purpose: the ear-bones of the haddock are, I think, the most suitable for this novel fancy-work.

To add a little varied colour to my screen, I embroidered a few butterflies, copied from nature, in coloured silks and introduced them with good effect amongst the jasmine sprays.

The screen is made up with old gold cord and tassels, and lined with silk of the same colour.

Scale Insects

I have made an acquaintance more curious than agreeable, in the shape of the destructive orange-scale insect. I find it constantly appearing upon the stem and leaves of a small seedling orange tree which I have been growing from a pip. Every few weeks brown oval scales have to be scraped off the small tree, else its health would be impaired, for these apparently insignificant things are really live creatures, each of them possessing six minute legs and a kind of beak with which it bores into the stem and sucks the sap of the plant. These scale insects are a serious annoyance to gardeners, and give rise to an immense amount of trouble, for they multiply rapidly, and when once a plant is infested with them there seems no remedy but washing carefully each individual leaf, or else syringing the entire plant with some poisonous liquid.

The life-history of the various scale insects is not fully known, but in most cases the male insect is a minute fly; the small tortoise-like brown atom which adheres to the stem and leaves being the female.

PALM SCALE. There are many English species, and unfortunately in importing foreign plants we are apt also to import new kinds of scale insects which find a congenial home in our hothouses. The palm scale is one of the most conspicuous, and if we remove one of these from the under side of a palm leaf in autumn we may, with a lens, discern about fifty white eggs within the brown shell, left there by the dead scale insect, ready to hatch in due time and perpetuate her species. On the fruit of both oranges and apples we may often find the mussel scales (*Aspidiotus conchiformis*). At the first glance we should take them to be mere brown specks, but the exact form of the mussel shell shows that they are true scale insects. As long ago as the year 1518 a kind of scale was observed upon cactus plants in Mexico (*Cossus cacti*). It was found to contain a red colouring matter, which forms the basis of the rich carmine used by water-colour painters, and it also

COCHINEAL INSECT ON CACTUS.

yields the cochineal of commerce, so much employed in dyeing and in various arts.

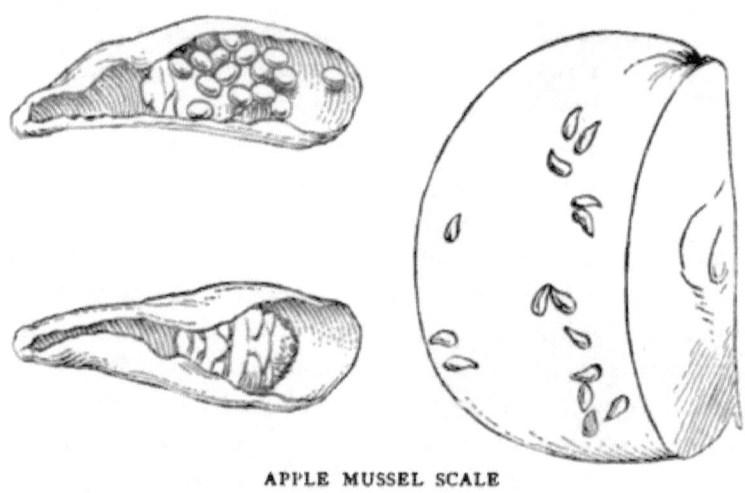

APPLE MUSSEL SCALE
(*Much magnified*).

An Indian scale insect (*Cossus lacca*) deposits a reddish waxy substance upon the twigs and branches of trees: this substance is called stick-lac, and is largely used in the manufacture of sealing-wax and varnish. So that while we look with disfavour upon the insect plagues of this species which infest our greenhouses, we may at least recollect that they possess foreign relatives who have a certain claim upon our gratitude.

Spiders

A spider's web empearled by hoar frost is indeed 'a thing of beauty.' To-day every tree branch, bush, and spray, is seen to be hung with these jewelled webs, even the lawn is covered with them, and one realises that flies live in a very world of snares unseen by us until the frost reveals them. There really seems to be a spider fitted to every situation in life.

In our houses reside the *tegenarias*, those black, long-legged, swift running creatures which are the *bêtes noires* of nervous people, but which, notwithstanding, are full of curious ways and instincts, as I can vouch for, seeing I kept one as a 'pet' for more than a year.

I watched with interest its making silken tunnels, laying its bag of eggs in a corner of the box it resided in, and concealing it by sticking all over it the legs and wings of the flies it had fed upon, until the egg nursery looked only like a bit of old spider's web.

Another species of spider haunts our window sills. It makes no web, but catches flies by lying in wait and springing suddenly upon them;

it is called zebra, from its lovely stripes and markings.

In pine trees we may find a spider of the most vivid green colour weaving small webs to entrap flies, and in some hidden corner it places a little mass of brilliant yellow silk which contains its precious store of eggs.

On the surface of ponds spiders may be seen running swiftly to and fro. One species elects to reside upon a floating leaf, and on this little raft it must lead rather a precarious life driven about with every gust of wind.

The most curious of the aquatic spiders is the one which dives below the surface, carrying with it a supply of air, with which it fills a silken bag it has woven amongst the weeds growing at the bottom of the pond. In this small balloon it lives its hidden life, preying upon small water insects, only going up to the surface now and again in order to renew its supply of needful air.

In summer we may see thousands of dark brown wolf spiders, each carrying a snow-white ball of eggs beneath its body, as it threads its way amongst the grass stubble where the hay has been cut and carried.

Even the air has its tenants from this ubiquitous tribe, for in autumn we may often see the tiny gossamer spider being wafted along with its trail of silken web floating past in the soft breeze.

All these creatures doubtless have their uses, and each performs some needful part in the economy of nature.

LEPISMÆ

There is a tiny dweller in our houses, not often seen, because of its nocturnal habits, but yet for several reasons it is worth a little careful study.

I paid a visit to my kitchen hearth last night when the lights had been put out and all was quiet. There I saw small silvery creatures, shaped like fishes, flitting rapidly about within the kitchen fender. These were *Lepismæ*, but when I endeavoured

LEPISMÆ SACCHARINA
(*Magnified*).

to catch them I found it by no means an easy task.

I managed it at last by means of a small dusting brush and a basin. With a rapid sweep of the brush I secured a few specimens, which I felt could only be safely retained in a glass globe, their small size and agility enabling them to escape from almost any kind of box.

When I examined them by daylight I saw that these singular little atoms possess six legs, two antennæ and three long hairs in the tail. They glisten as if formed of silver, and their scales are so fine and delicate as to be used as a test for microscopic glasses.

I have kept *Lepismæ* for months, feeding them on cake and sugar until they became tame enough to bear being looked at without fear. The Latin name, *Lepisma saccharina*, implies their preference for sugar, although they indulge in other rather diverse articles of diet, such as sweet cake, wall-paper, book bindings and furniture coverings.

They are often to be seen in damp libraries running over books and papers, but they are so small that I do not think much injury can be laid to their charge.

The Germans call these little creatures silver fishes, a name which accurately describes their appearance.

THE CHRISTMAS ROSE (*Helleborus niger*)

The Christmas rose, which has come to cheer us with its snow-white flowers, is an imported plant from Southern Europe. Two species of hellebore are, however, found growing wild in some parts of England, though even they are not believed to be truly indigenous.

Helleborus fœtidus is now flowering in my garden, and is an interesting and rather showy plant, with clusters of green bell-shaped flowers edged with purple.

Helleborus viridis is found on chalky soils, and has also pale green flowers and dark green leaves.

The species figured in the illustration is *Helleborus*

purpurascens; it shows very plainly the curious construction which is common to all hellebore flowers. What we should naturally call the petals are really the leaves of the calyx—called sepals—which do not fall off, but after a time become of a greenish hue, and share in the work of leaves by helping to nourish the plant. Instead of petals we find tubular nectaries filled with honey, which are situated between

PURPLE HELLEBORE
(*Showing honey-glands*).

the sepals and the stamens. These tubes are attractive to bees from the sweet though poisonous liquid they contain, and in thus rifling the nectaries they brush the pollen on to the stigma and fertilise the flower.

It seems strange that the Christmas rose with its snowy flowers should be called black hellebore, but it is so named from its dark root-stock and black fibres.

December

Mites

Some valuable foreign insects in my museum have been reduced to a heap of dust by an army of microscopic mites, whose life work it is to demolish dried specimens, and whether they are butterflies, wasps, beetles, or plants seems immaterial to them.

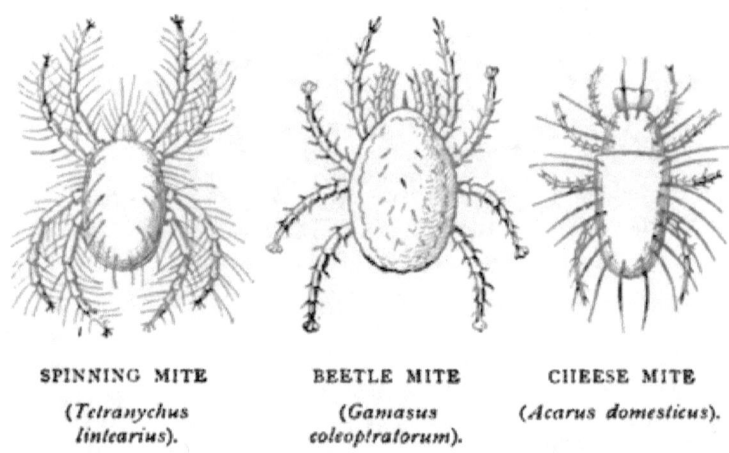

| SPINNING MITE | BEETLE MITE | CHEESE MITE |
| *(Tetranychus linteariıs).* | *(Gamasus coleoptratorum).* | *(Acarus domesticus).* |

This incident has led me to some slight study of the mite family; and I am surprised to find how many species there are, and what widely differing kinds of work they are engaged upon.

We all know the cheese mite, which quickly reduces our favourite Stilton to a mass of powder; this much resembles the destroyer of dried butterflies, and both are like a certain other mite which abounds in damaged flour.

There is a special mite which eats dried figs; another species prefers dried plums.

The feathers of the ostrich are infested by a minute creature of this kind, and it is also found in owls' plumage.

In the cavities of the bones of skeletons mites exist, and old honeycomb is quickly taken in hand by them and destroyed. A specimen of the sacred beetle of Egypt was sent to me alive some years ago. I kept it in health for about sixteen months, but so rapidly did mites breed upon its living body that every few weeks I had to place it in warm water, and with a camel's hair pencil brush away dozens of minute specks which I could only just discern running over its body.

Sometimes humble bees are infested in this way, and I pick them up in a dying state, apparently unable to rid themselves of their tormentors.

The excessive irritation many persons experience after walking in cornfields is due to the harvest-mite, which buries itself in the skin and there creates acute imflammation and much consequent distress.

Some years ago I met with another branch of the family, and could but marvel at its extraordinary labours. A furze bush was apparently wreathed in fine white muslin in layers between the branches, fold after fold, and upon this gauzy material were multitudes of bright red specks careering about. Of course I took some specimens home, and I soon discovered they were spinning mites (*Tetranychus lintearius*).

There are many species, and it is one of these, the so-called 'red-spider,' which does so much mischief in greenhouses by sucking the juices of plants.

Birds are sadly worried by a small red mite, which lives in the crevices of cages which are not kept perfectly clean. The best protection from their attacks is a good sponging of the perches and every part of the cage with a solution of carbolic acid; this will effectually get rid of the insects.

I am not attempting to write an essay upon mites, or else I might speak of dozens of other species, some parasitic upon flies and spiders, and others inhabiting ponds and ditches. I have but touched upon a few kinds I have happened to meet with in daily life.

The minute creatures evidently have an appointed work, which they do secretly and mysteriously, all unknown to us, until a suspicious heap of dusty fragments shows where this unseen army have been encamped.

Birds' Heads

Having drawn attention to the feet of birds, as affording a clue to the kind of life to which they are adapted, I will now try to show how the formation of a bird's skull and beak indicates the character of the bird, and the kind of food it lives upon.

Ornithology is a very wide subject, seeing that there are said to be over ten thousand species of birds. These are grouped into about twenty-two orders, and of these I have selected four specimens to illustrate my remarks.

An owl's skull, with its curved and sharply-pointed beak reveals the fact that the bird is a flesh-eater, catching its prey alive. One grip of an owl's claws suffices to kill the captured mouse or bird, and then the beak tears the prey into fragments. As a matter

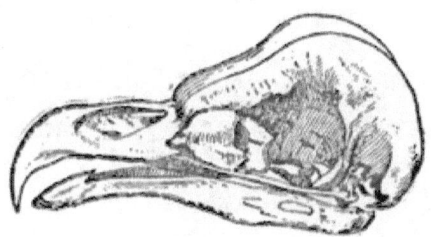

OWL'S SKULL.

of fact the owl swallows mice whole, but when it is kept in a cage and supplied with raw meat, we see the powerful beak tearing the flesh to pieces, just as an eagle would dismember its living prey.

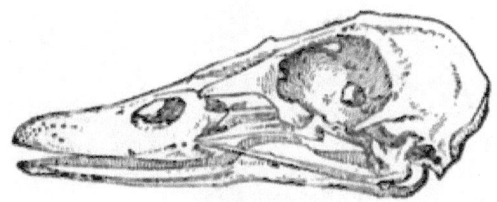

DUCK'S SKULL.

The broad, spoon-shaped beak of the duck has a lining of horny ridges, which enables the bird to mince and prepare its vegetable and fish diet.

In the skull of the woodcock we may observe that the eye orbit is placed far back, so that the beak may be plunged up to its base in soft mud, where it feels about for the worms on which the bird subsists.

The pheasant is a type of a grain-eating bird, the bill being short and powerful, much like that of the common fowl.

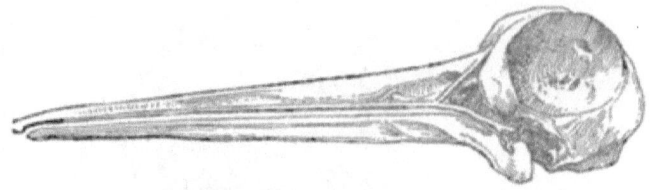

WOODCOCK'S SKULL.

It is well to know the difference between what are called hard-billed and soft-billed birds, since we may desire to bring up some fledgeling, and be in doubt as to the food suitable for its needs. If it has a beak like the common sparrow, then it is

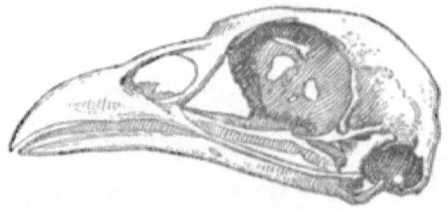

PHEASANT'S SKULL.

hard-billed, and a grain-eating bird. We may be sure it will thrive upon some such food as sopped brown bread, or a paste made of oatmeal and water. If, on the contrary, the beak is slender, like a robin's, it is called soft-billed, which is an indication that it is an insect-eater. Such young birds must have a diet of finely minced raw beef mixed with a little sopped brown bread. If my readers will notice, either in pictures or at the Zoo, the endless variety of form in the beaks of birds, English and foreign, they will admire the marvellous adaptation to the needs of each bird, from the great pouch of the fish-eating pelican, down to the slender curved beak which enables the humming-bird to obtain honey from tubular blossoms in the Tropics.

Index

	PAGE		PAGE
Abele	60	*Bibio*	69
Abraham's Oak	144	Birds, feet of	185
Acanthus	158	,, heads of	215
Acherontia atropos	71	Bird's-foot trefoil	134
Acorns	176	Blue Anchor, alabaster at	142
Acrida viridissima	133	Bluebottle	86
Adams, H. G., quoted	48	Blue-tits	148
Agaricini	153	*Boletus edulis*	156
Alabaster	142	*Bombylius major*	74
Alder catkins	42	Broad-bean	111
Alnus glutinosa	42	Brontë, Charlotte, quoted	116
Amadou	156	Broom-rape	160
Amber	117	Bryant quoted	18, 19, 132
Aphrophora spumaria	95	Butterflies	138, 151
Aquatic spiders	209	Buxton, fossils at	141
Araucaria	59		
Argonaut	193	Callimachus	158
Ash	75	Campanula	181
Ash-bark beetle	26	Canary grass	87
Aspen	60	*Cannabis sativa*	147
Aspidiotus conchiformis	206	*Capsella-bursa pastoris*	79
Aucuba	49	Capsules of flowers	178
		Caterpillar dwelling	196
Balsam poplar	60	Catkins, tree 40, 60, 63, 91,	177
Bats	149	Caverns	141
Bede quoted	107	Cedar of Lebanon	176
Bee, nest of, in snail shell	195	*Cedrus*	176
Beech, catkins of	92	Charlock	136
,, seed-leaves of	79	Cheddar Cliffs, cavern at	141
Bees	45, 99, 195	Cheese mites	213

Index

	PAGE		PAGE
Chicory	127	Flies killed by fungus	190
Christmas rose	211	Flints	24, 119
Cichorium	125	Floral clock	126
Clare quoted	30, 95	Flowering trees	90
Clavarias	157	Flowers, capsules of	178
Claws of birds	185	Footprints in snow	36
Cloth moth	32	Fossils	24, 141
Coachman fly	70	Frog-fly	96
Cochineal insect	207	Frog-hoppers	95
Compositæ	180	Frog-spit	95
Coprinus comatus	136	Frost, effects of	33
Corinthian order, origin of	158	Fuller's teasel	123
Cork moth	32	Fungi	153
Cossus	207	Fur and feather moth	32
Cow parsnip	123		
Cuckoo-fly	96		
Cuckoo-spit	95	Galls	169
Cyclamen	178	Gladiolus	84
Cynips	169	Goethe quoted	65
		Gossamer spider	209
Dandelion	54	Grain-eating birds	187
Datura	179	Granite	120
Death's head moth	71	Grasshoppers	133
Derbyshire caverns	141	Great green grasshopper	133
Dipsacus	122	Grey poplar	60
Draba verna	52	Growing seeds	110
Dragon-fly	128		
Dry rot	156		
Duck skull	216	Hard-billed birds	217
Dyer quoted	124	Heads of birds	215
Dytiscus marginalis	128	Heartsease	180
		Helix lapicida	194
Electron	118	Hellebore	211
Empusa musci	190	Hemp	147
Endive	127	Henslow, Prof. G., quoted	162
Exoascus carpini	23	Hidden lives	195
		Hieroglyph for joy	140
Felspar	122	Holly	201
Finches	136	Holmes, Dr. O. W., quoted	192
Flax	105	Homer quoted	118
Flies in Amber	118	Hornbeam	23, 59, 179

Index

	PAGE		PAGE
Horse-chestnut	22	Long-eared bat	149
House fly	85	*Lotus corniculatus*	134
Hoverer-fly	102	Luther, Martin, quoted	139
Humble-bee fly	74		
Husbandman's tree	77		
Hydnum repandum	157	Maned agaric	136
		Mant, Bishop, quoted	75
		Maple	59, 179
Ichneumon flies	103	Mason bee	99
Ink, home-made	136	Mason wasp	196
Insects and flowers	83	Mealworm beetle	50
Iris	179	*Medicago helix*	194
Ivy-leaved toad-flax	92	*Megachile centuncularis*	101
		Merulius lacrymans	156
		Milton quoted	119
Jacana	186	Mimicry of seeds	194
Jasper, red	120	Mites	213
		Montgomery quoted	193
		Moorhen	186
Keats quoted	164	Morris, Wm., quoted	98
		Moths, minute	31, 198
		Musca domestica	85
Lamorna	121	Mushrooms	153
Larch seedlings	90	Mussel scale insects	206
Larder fly	87	*Mycelium*	154
Lark	188		
Laurel-leaf glands	50		
Laurus nobilis	50	Nautilus	191
Leaf-cutter bee	99	Nightingale	52
Leaf-scars	22	*Notonecta glauca*	127
Leaves, positions of	139	Nuthatch	55
Lemon pips	112		
Lepisma	210		
Lesser celandine	38	Oak, galls of	170
Leucadendron argenteum	180	Oaks	142
Liberation of seeds	178	,, English	173
Lime tree seed-leaves	79	*Oinophila v. flava*	31
Linaria cymbalaria	92	Oolite	142
Linum catharticum	108	Orange pips	112
Linum usitatissimum	105	*Orobanche*	160
Lombardy poplar	60	Orthoclase	122

Osmia	99, 195
Otoliths	202
Owls	20, 185, 216
Palestine oaks	142
Palm scale	206
Paper, home-made	167
Papyrus	165
Patmore, Coventry, quoted	149
Pezizas	157
Phalaris Canariensis	87
Pheasant skull	217
Phyllotaxis	22
Pigeons, wild	136
Pimpernel	181
Pinus succinifer	117
Plumule	111
Podophyllum	113
Polygonums	136
Polyporei	155
Pope quoted	192
Poplar catkins	60
Poppy	181
Populus	60
Pratt, Miss A., quoted	94
Ptarmigan	186
Pteromalus	19
Pudding-stone	120
Pulvinus	139
Quercus	144, 174
Ranunculus ficaria	38
Raphidia ophiopsis	108
Raptorial birds	185
Red spider	214
Rhodites rosæ	169
Rhododendron	83, 140
Rhyssa persuasoria	104
Robin's pincushion	169
Rocks	119, 141
Roe-stone	142
Rossetti quoted	48, 68
Rowden, F. A., quoted	62
Royal Natural History quoted	70
St. John's fly	69
St. Mark's fly	69
Salicine	66
Samara	58
Sanguinaria	113
Sarcophaga carnaria	87
Saxifraga tridactylites	54
Scale insects	205
Scolytus	26
Screen	204
Sedum, butterflies in	151
Seed mimicry	194
Seedling trees	77
Seeds, growing	110
,, liberation of	178
Shelley quoted	200
Shepherd's purse	79
Silver fishes	211
,, tree	180
Sitta Europæa	55
Skeleton leaves	35
Skulls of birds	215
Snake-fly	108
Snapdragon	181
Snow crystals	33
Soft-billed birds	217
Southey quoted	201
Spangle-galls	172
Spiders	208
Sponges, fossil	24
Squirrels	173
Stalactites	141
Stalagmites	141
Stick-lac	208
Stonecrop	152

Index

	PAGE		PAGE
Stones	119, 141	Vespa	43
Succory	125	Voles	21
Swimming birds	186		
Sycamore	58, 78, 179	Wasp, mason	196
Syrian oaks	144	Wasps	43
Syrphus plumosus	102	Water-boatmen	127
		Water-lion	129
Tacamahac	60	Water-rail	186
Tamarind seeds	112	Weeds, use of	136
Teasel	122	White flax	108
Tegenarias	208	White poplar	60
Tenebris molitor	50	Whitlow grass	52
Tennyson quoted	62, 75, 82, 158	Wild balsam	179
Tetranychus lintearius	214	,, geranium	180
Thomson quoted	184	,, pimpernel	180
Tineas	31, 196	,, succory	125
Titmice	148	,, teasel	122
Toadstools	155	Willow catkins	41, 63
Tree catkins	40	Witches' brooms	23
,, seeds	57	Wolf spiders	209
Trillium	88	Wood, Rev. J. G., quoted	73
Trinity flower	88	Woodcock skull	216
Trumpeter pigeons	187	Woodsorrel	180
Tsetse fly	71	Wood wasp	45
		Wordsworth quoted	38, 146
Upholsterer bee	101		
		Yew	62
Valonia oaks	144		
Ventriculites	24	Zebra spider	208

Printed by Hazell, Watson & Viney, Ld., London and Aylesbury.

www.ingramcontent.com/pod-product-compliance
Lightning Source LLC
Chambersburg PA
CBHW021843230426
43669CB00008B/1066